The Mathematical Universe

From Pythagoras to Planck

Joel L. Schiff

The Mathematical Universe

From Pythagoras to Planck

 Springer

Published in association with
Praxis Publishing
Chichester, UK

Joel L. Schiff
Mairangi Bay, New Zealand

SPRINGER-PRAXIS BOOKS IN POPULAR SCIENCE

Springer Praxis Books
Popular Science
ISBN 978-3-030-50651-3   ISBN 978-3-030-50649-0   (eBook)
https://doi.org/10.1007/978-3-030-50649-0

Project Editor: Michael D. Shayler

This Springer imprint is published by the registered company Springer Nature Switzerland AG
The registered company address is: Gewerbestrasse 11, 6330 Cham, Switzerland

Contents

Dedication viii

Acknowledgements x

About the Author xi

Foreword xii

Prologue xiv

Preface xvii

1   The Mystery of Mathematics 1
   Let us be reasonable 3
   All set 7
   Where is Mathematics? 9
   Fine tuning 12
   A blast from the past: Euclid's geometry 17
   Taking the Fifth further 20
   Pi in the sky 28
   Off to Monte Carlo 30
   Smashed pi 32
   The divine Isoperimetric Inequality 35

2   From Here to Infinity 39
   Zeno's Paradox 41
   Summing up 52
   In what Universe is this true? 60
   The power of e 67
   Fast money 71
   What is normal? 72
   Multiplying *ad infinitum* 74

3   Imaginary Worlds   76
The Strange Case of $x^2 + 1 = 0$ 77
The 'i's have it.   80
The God-like Euler identity   82
Even more imaginaries – quaternions.   86
But wait, there is more – octonians.   89
The world's hardest problem – the Riemann Hypothesis   91

4   Random Universe   99
Going steady   99
Brownian Motion   103
Life is a gamble.   103
Exponential decay.   105
The dating game   106
Empowering laws   109
The world of entropy – order to chaos   113
Information entropy   116

5   Order from Chaos   120
Cellular Automata   120
Life as a game   127
Infectious disease model – SIR.   131
Mimicking Darwin   132
One-dimensional CA   137
The whole is greater than sum of its parts   142
Bees and termites   143
… And ants   144
Bacteria count   147
A hive of Mathematics: Fibonacci   149
Dynamical systems   152
Messrs. Fatou, Julia, and Mandelbrot   157
The fractal Universe   162

6   Mathematics in Space   168
Faster than a speeding bullet.   168
Down to Earth.   173
Heavens above   175
Light-years   178
The great recession   180
The Universe is flat   185
Measuring the invisible: Black holes   188
A galaxy far, far, away   192

7   The Unreality of Reality   194
Miniature Universe   195
Quantum world   199

Infinite space 206
Qubits 209
It is all relative, Albert. 213
That equation 217
What time is it anyway? 218
Matters of gravity 219
Time in motion 228
Radiation. 234
Symmetry and groups 236

8   The Unknowable Universe 245
Gödel incompleteness. 245
Halting problem 246
EMX 247
Where is it, Dr. Heisenberg?. 248
Summing up 250

Appendix I   Being Reasonable 253
Appendix II   Hyperbolic Geometry and Minkowski Spacetime 256
Appendix III   The Uncountable Real Numbers. 259
Appendix IV $c^2 = c$: Square and Line have Same Cardinality 261
Appendix V   Geometric Series. 263
Appendix VI   Cesàro Sums 265
Appendix VII   Rotating a Vector via a Quaternion. 266
Appendix VIII   Quaternions $q^2 = -1$ 269
Appendix IX   Riemann Zeta Function. 270
Appendix X   Random Walk Code 274
Appendix XI   Age of the Solar System 276
Appendix XII   Chelyabinsk Meteoroid. 279
Appendix XIII   Logic Gates. 280
Appendix XIV   Galaxy Distance via Cepheids 283
Appendix XV   Time Dilation 286
Appendix XVI   Expansion of the Universe 289

Bibliography 291

Index. 294

Not only is the universe stranger than we imagine, it is stranger than we can imagine...
Sir Arthur Eddington

This book is dedicated to the memory of Emmy Noether (1882–1935), who made many brilliant contributions to the world of Mathematics and whose eponymous theorem changed the world of Physics forever.

Amalie Emmy Noether, date unknown but prior to entering Göttingen University in 1915. Renowned 20th century mathematician Norbert Wiener asserted: "Leaving all questions of sex aside, she is one of the ten or twelve leading mathematicians of the present generation in the entire world." (Image in public domain.)

Acknowledgements

This book could not have been written without the generous assistance, in various forms, of numerous people whom I sincerely wish to thank. In particular, I would like to thank copy editor Mike Shayler for significant improvements to the wording of the text. Any errors, however, can only be blamed on the author. The others I wish to thank are:

Scott Aaronson, University of Texas
John Baez, University of California, Riverside
Bonnie Bassler, Princeton University
Michael Berry, FRS, University of Bristol
Mario Bonk, University of California, Los Angeles
Douglas Bridges, University of Canterbury
Cristian Calude, University of Auckland
Richard Easther, University of Auckland
Cohl Furey, University of Cambridge
Courtney Griffith, University of Santa Clara
Gerard 't Hooft, Nobel Prize in Physics 1999, University of Utrecht
Vaughan F.R. Jones, FRS, Fields Medal 1990, University of California, Berkeley
Anita Kean, NeSI/University of Auckland
Chris King, University of Auckland
Ron King, University of Southampton
Kevin McKeegan, University of California, Los Angeles
Katy Metcalf, Auckland, for the many fine illustrations
David Morrison, NASA/ARC
Aaron Schiff, Economist/Data Scientist, Auckland
Ryan Schiff, Lockheed Martin
Scott Tremaine, Institute for Advanced Study, Princeton
Andrew Yee, HPS Specialist
Philip Yock, University of Auckland

About the Author

Joel L. Schiff has a PhD in Mathematics from the University of California Los Angeles (UCLA). He has spent his career at the University of Auckland, Auckland, New Zealand and has written five books on mathematical and scientific subjects. With colleague Wayne Walker, he helped develop the *Arithmetic Fourier Transform* used in signal processing. He was also the founder publisher of the international journal *Meteorite* and in 1999, he and his wife discovered a new asteroid from their backyard observatory. They named it after the notable New Zealand meteorite scientist, Brian Mason.

Foreword

On a sunny day, one million years ago, our world already looked much like it does now. There were mountains, forests, and deserts, and they were inhabited by plants and animals, much like it is today. But something very special was happening among some small groups of primates. They had grown to become more intelligent than any other animal, and they noticed all sorts of remarkable features in the world they lived in. In the weather, in the fields, and in the sky, they saw patterns, varying during the days and nights, and during the seasons.

They learned how to communicate among one another, how to foresee what was going to happen, and how to use their insights to improve their safety and to ensure the availability of food and shelter. This is how science began, and the primates could no longer be regarded as animals; they started to talk; they were becoming human.

Then, a few thousand years ago, they began to write and read. They learned to use new concepts of thought, such as numbers. They started to use their fingers to sort out the different numbers that one might have and soon discovered that we don't have enough fingers to account for all possible numbers. You can go all the way to 60, but then it gets harder. You have to combine numbers to describe larger numbers, and sometimes you need numbers that aren't integers to indicate how much water you have or how tall something is. This was the beginning of calculus. You can also use numbers to register time, like the hour of the day or the day of the year, and to get this accurately required some real thinking.

It sometimes happened that they thought they understood everything that was to be known, but time and again this turned out to be wrong. The art of using numbers to indicate the size of things became known as mathematics, and the science of describing forces and other features of the things you see became known as physics. Studying physics and mathematics became a specialty that only a few experts can understand. This led to the need for people who could bridge the gap between the experts and the rest of the population. It is not an easy job, but I think this book is a delightful endeavor in this regard.

Gerard 't Hooft (Nobel Prize in Physics, 1999).

Prologue

A BRIEF TALE

Early in the 20th century, a patent office clerk (class III) in Bern, Switzerland, was thinking about light. It was a topic he had been dwelling upon since he was 16. His patent office work was not very demanding, so he had time to think on his own. As a starting point, he stipulated two basic postulates: one about the constancy of the speed of light in empty space, the other about the invariance of the laws of Physics.

The clerk then considered the ramifications of these two principles. He toiled away, without doing any physical experiments of his own, but rather experiments inside his own head; that is, 'thought experiments'. Simply from his deliberations, a number of equations flowed from his pen in due course.

One of the conclusions drawn from the mathematical squiggles that the clerk wrote down was that a clock aboard a moving spaceship would tick more slowly than one back on Earth, and thus time would actually slow down for the space travelers. A year in space for them could even correspond to ten years back on Earth.

Yes, *time*, which had always been thought to be an immutable quantity, was no longer so. At least, that is what the equations said.

After the publication of the clerk's ideas, he turned his thoughts to gravity, another notion we are all familiar with, and the clerk eventually became a university professor. After some ten years of deliberation on gravity, his new squiggles showed that gravity could also slow down time, and could warp space that bent light beams which had no mass.

Many found this state of affairs very confusing and few could comprehend the meaning of the squiggles. This was time and space after all, things people were intimately familiar with from a lifetime of experience. The clerk's revelations simply defied common sense.

The above account has not been taken from some bizarre science fiction story. These considerations concerning light and gravity were spelled out in two landmark publications concerning the Theory of Relativity, in 1905 and 1916, respectively. The author's name was Albert Einstein. His squiggles on paper completely redefined the nature of time and space, and his theories have proven to be a profound platform for understanding the workings of the Universe. All the theoretical results of his work have been scientifically verified, often to many decimal places.

Another fantastical prediction to come out of Einstein's squiggles was the notion of black holes.

Albert Einstein, painted by American artist Paul Meltsner (1905–1966). This painting was responsible for the purchase of $US 1 million in war bonds in 1943. (Image courtesy of Rabbi Naomi Levy/Rob Eshman. Photo by Ted Levy.)

But do seemingly abstruse mathematical equations have anything to do with our everyday lives? The answer is yes. For without a complete understanding of how time can slow down and speed up aboard an orbiting satellite, there would be no Global Positioning System (GPS) and everyone would become hopelessly lost traveling from A to B. Well, at least some of us would and many of us did. Moreover, GPS is an integral part of aircraft navigation systems, as well as that of ships at sea.

Other squiggles that appeared in Einstein's first publication on Relativity said that mass and energy were equivalent, obeying the relation: $E = mc^2$. Now these three letters, plus one number and an equal sign, arranged just so, have had a most dramatic effect on the course of human history. Moreover, it is also operating every second of every day in our Sun, converting mass to energy and bringing us light. The light with which to explore the nature of reality all the way back to the origins of our Universe.

Preface

Nature imitates Mathematics...
Mathematician Gian-Carlo Rota

This is a book about the mathematical nature of the Universe we live in. Perhaps in some other universe, if such exists, the Mathematics would be different. The question of whether our Mathematics is discovered or invented has exercised many great minds since the time of Plato (who claimed it was discovered). We will touch upon this issue since the author has his own view of the matter, as most mathematicians do. Regardless, there are certain fundamental features of our Universe, and even predictions that are beautifully described by the little squiggles on paper that we call Mathematics.

What the author aims to do in the pages that follow is to explain some of the mathematical aspects of the Universe we live in by using no more than basic high school Mathematics. What is explored through these chapters are some selected topics that are easily accessible to anyone who has once brushed up against the Pythagorean theorem and the symbol π, with a light dusting of algebra. But do not worry if you have forgotten the details as these will be provided. As a bonus, nothing involving even a smattering of Calculus is needed for any of our considerations.

We will make numerous stops along the way, visiting the sublime but bizarre world of subatomic particles, honey bees, the Theory of Relativity, galaxies, black holes, and of course infinity, as well as a universe of our own creation with its own set of rules called Cellular Automata. The point is, the whole Universe is linked by bits of Mathematics that most people already know, and what the reader does not already know, the author hopes that you will come to understand by the end of the book because you will find it compelling and accessible.

So, this is a Science book for the layperson that has a few equations, unlike many other such Science books that are purely descriptive. But simply knowing the equation for a straight line ($y = mx + b$) enables us to determine the age of our Solar System, or to measure the size of a supermassive black hole that lies at the heart of most galaxies, or even allows us to describe the expansion of the Universe itself. Moreover, we will describe how time speeds up and slows down and you will become a believer. Now all that is pretty extraordinary, and just a sample of what we can do with the most basic of squiggles.

We will only be able to touch on the highlights of our Mathematical Universe, since Mathematics is at the very core of so much scientific endeavor and the rigorous details of those endeavors require concepts that are beyond the scope of this book. Nevertheless, there is much we can explore with the simplest of mathematical tools.

The author cannot guarantee that you will have six-pack abs by the end of this book, but the mental exertion could possibly lead to a six-pack brain.

1

The Mystery of Mathematics

Pure Mathematics is religion...
Philosopher Friedrich von Hardenberg

It is impossible to be a mathematician without being a poet in soul...
Mathematician Sofia Kovalevskaya

Everyone who has gone to school has learned some Mathematics, with the experience for many being a painful one. Whenever someone on a plane or at a party asks the author what he does and he replies that he is a mathematician, the conversation either halts immediately, as if he had said he was an undertaker or worked for the Internal Revenue Service, or they confess that they were never very good at math at school. So, he appreciates the phobia, dread, and forbidding nature regarding what he is about to say, but can assure you that it will be entirely painless.

In 1999, the author published a book with the title *Normal Families*. The title was somewhat misleading, and he suspects that some copies of the book were purchased thinking that it held some deep psychological insights into family life. However, it was entirely devoted to a very esoteric branch of Mathematics by the same name. Indeed, it included many beautiful theorems that had absolutely no bearing on the natural world. In fact, most of the material was as remote from reality as it could be. Yet some of the results could even be considered majestic, in the same way that Mahler's Fifth Symphony might be so considered. The structure of their proofs was just so elaborate, so rich and magnificent in their construction, and the final results so illuminating of other dark corners of the mathematical

© Springer Nature Switzerland AG 2020
J. L. Schiff, *The Mathematical Universe*, Springer Praxis Books,
https://doi.org/10.1007/978-3-030-50649-0_1

realm, that the author was often in awe of the genius that went into their creation. These were the results of others and he could in no way take any of the credit for the results themselves. He was merely the messenger.

But that begs the question, 'Messenger of what'? What really is Mathematics? Where does it reside? Is it the product merely of our imagination, or is it found in some netherworld outside of space and time. Is Mathematics a religion? Is it the language of God? Or even, is God Mathematics?

These are very deep questions in and of themselves and have been fretted over for millennia by mathematicians and philosophers alike. The author has pondered over them himself for the more than four decades that he has been a mathematician, since he has frequently wondered what in the world he has actually been doing all these years. It is hoped, however, that by the end of this book, the reader will have a better understanding of what Mathematics is, although there is no simple answer.

There is at least a 4,000-year glorious history of Mathematics and the sophistication of some of the earliest work is quite remarkable. For example, the Babylonians used a base 60 system of numbers. Indeed, we still do when measuring seconds and minutes of time, or in arcseconds and arcminutes of angle. The diagonal of the square in Fig. 1.1 (with the horizontal row of numbers) represents:

$$1 + 24/60 + 51/(60)^2 + 10/(60)^3 = 1.41421296,$$

which gives the value of $\sqrt{2}$ accurate to 6 parts in 10 million $\left(\sqrt{2} = 1.41421356...\right)$.

The square root of 2 was known mathematically as the ratio of the diagonal of a square to a side of length 1. In this particular ancient school exercise, the value 1 is replaced by the value '30' at the top left for the length of a side, so that the diagonal would have a length 30 times greater, namely: $30 \times \sqrt{2} = 42 + 25/60 + 35/(60)^2$, represented by the bottom numbers. This is sophisticated mathematics beyond the capability of any measuring device at the time, and would (using base 60) be a challenging problem for a high school student of today. Try it.

However, to do the historical side of Mathematics justice would require a completely separate volume to this one. Nevertheless, the names of many famous individuals who made important contributions to our understanding of the mathematical and physical worlds are sprinkled throughout this text.

In the remainder of the text, we shall explore the mysterious relation between Mathematics and the Universe, for without Mathematics we would have little left to explore. We could not even count sheep at night to go to sleep. But first we need to consider some of the basic elements of mathematical logic in order to make this exploration possible.

Figure 1.1: They were not just counting goats and sheep in Babylonian times. This is the cuneiform Babylonian school tablet YBC 7289 from 1600−1800 B.C. See text for explanation. (Image courtesy of Bill Casselman (https://www.math.ubc.ca/~cass/Euclid/ybc/ybc.html) and Yale Babylonian Collection.)

LET US BE REASONABLE

> *Logic: The art of thinking and reasoning in strict accordance with the*
> *limitations and incapacities of the human misunderstanding...*
> Ambrose Bierce, *The Devil's Dictionary*

Mathematics at its heart relies on the power of reasoning in a rigorous fashion. Such systematic reasoning, known as *symbolic logic*, is a mode of thought that was initiated by Aristotle, developed by the Stoics, and further expounded upon in a more mathematical setting beginning in the 19th century by George Boole, Augustus De Morgan, and Charles Sanders Peirce, among others. It is an attempt to make reasoning − and in particular mathematical reasoning − highly rigorous.

We use basic forms of logical reasoning all the time in our daily lives without even realizing it:

If it is 2:30, it is time to go to the dentist[1].
It is 2:30.
Therefore, it is time to go to the dentist.

This sort of reasoning, or *inference rule*, has a specific name: *modus ponens*. Another basic inference rule is known as *modus tollens*, as in:

If my grandmother had wheels she would be a trolley car[2].
My grandmother does not have wheels.
Therefore, she is not a trolley car.

Both of these forms of logical inference have their origins in the mists of antiquity. In classical logical reasoning, any logical statement (proposition) P in the form: 'It is time to go to the dentist' (or 'my grandmother is a trolley car') is considered to be either true or false. The *negation* of P is the statement: 'It is not time to go to the dentist' (or 'my grandmother is not a trolley car') and is referred to as the statement '*not P*', as in: 'it is *not* the case that P is true'.

Another form of logical reasoning going back to Aristotle is that:

Either a statement, P, is true, or its negation, not P, is true.

Thus, 'it is 2:30' or 'it is not 2:30'; either 'my grandmother is a trolley car' or 'my grandmother is not a trolley car'. There is no middle ground and that is why this mode of thinking is called the *law of the excluded middle*.

This is a cornerstone of mathematical reasoning, whereby any given mathematical statement is either true or false. Either $17 + 32 - 6 = 43$, or it does not; either 10,357 is a prime number or it is not[3]. If we have the statements (a) 10,357 is a prime number and (b) 10,357 is not a prime number, then one or the other must be true.

Just in case you were wondering, 10,357 *is* a prime number, so that it is only divisible by the number 1 and itself.

---

[1] To make the time more explicit, it should have been given as 14:30, but there is an implied child's joke here and dentists generally do not work at 2:30 am anyway.

[2] Trolley car is the American term for tram car. The expression, which the author heard many times as a young child, is a response to someone who makes a wistful statement of pure speculation, such as, "If only my parents had made me persist with my violin lessons, I could have become a great violinist."

[3] Recall that a prime number is a positive integer (counting number) greater than 1 that cannot be divided (with no remainder) by any other positive integer except for the number 1 and itself. Thus, the first prime numbers are: 2, 3, 5, 7, 11, 13, 17, 19, 23, ...

Based on the law of the excluded middle, here is an argument that many of us have no doubt encountered on the school playground. Suppose our proposition *P* is:

P: There is no largest number.

The negation of *P* is therefore simply the statement:

not P: There *is* a largest number.

If we assume for a moment that *not P* is the true statement, let us call the largest number *N*. But then *N* + 1 is still larger, so that the statement, *not P*, cannot be true. Since one of the statements *P* or *not P* must be true according to the law of the excluded middle, it follows that our proposition *P*: 'there is no largest number', is the true statement.

Indeed, with this playground example we have actually proved a genuine mathematical theorem, namely that *there is no largest number*. If the reader at some stage has enunciated some form of the above proof, it likely would have been done without ever realizing you were using sophisticated forms of logical reasoning and doing something mathematicians do every day, which is to prove theorems.

As seemingly obvious as the *law of the excluded middle* appears, it came under attack from a 20th century Dutch mathematician named L.E.J. (Luitzen Egbertus Jan) Brouwer (1881–1966), who rejected it on philosophical grounds when dealing with infinite sets. For Brouwer, an infinite set is something that is incomplete, as for example, the natural numbers, 0, 1, 2, 3, … which cannot be thought of in their entirety no matter how smart you are. That is why we need the three dots (ellipsis), which means *and so forth in the same vein*. See Appendix I for further discussion on this matter, where we give a proof by contradiction, as well as a 'constructive' proof of the proposition:

P: There is an infinite number of primes.

In 1946, the famous mathematician Hermann Weyl wrote, "Brouwer made it clear, as I think beyond any doubt, that there is no evidence supporting the belief in the existential character of the totality of all natural numbers... The sequence of numbers which grows beyond any stage already reached by passing to the next number, is a manifold of possibilities open towards infinity; it remains forever in the status of creation but is not a closed realm of things existing in themselves."

Indeed, there is a world of difference between the realm of finite entities and infinite ones and just how we treat the infinite is the subject of Chapter 2.

On the other hand, the law of the excluded middle leads to a method of proof known as 'proof by contradiction', which sounds more impressive in Latin: *reductio ad absurdum*. This is just the line of reasoning we employed to prove our little theorem that there is no largest number. We assumed the negation of the theorem to be true, i.e. that there is a largest number, *N*, and derived a contradiction since

the number $N + 1$ is obviously larger. Therefore, the negation of the theorem cannot be true and it follows that our theorem must be true after all: There is no largest
number. This is actually a very powerful technique when it is not possible to find
a direct proof of a theorem. In Appendix III, there appears a famous proof by
contradiction.

The subject of logical reasoning is a fascinating one but takes us too far afield,
so the interested reader is directed to the excellent book by R.L. Epstein in the
Bibliography.

Interestingly, some of the very basics of symbolic logic find their way into the
design of electrical switches, known as logic gates, that are at the heart of electronic computers. See Appendix XIII, where the three most fundamental logic
gates are discussed in relation to symbolic logic.

It should also be mentioned that Brouwer is perhaps more famous for his 'fixed-
point theorem.' A simple example would be to take a circular disk and rotate the
disk 30° about the center. At the end of the rotation, which is a smooth continuous
action, all the points in the disk will have moved, except for one point, the center,
which remains fixed. Brouwer's theorem says that any continuous action that
transforms a suitably closed region to itself will always leave at least one point in
its fixed position. Ironically, Brouwer's proof of his fixed-point theorem is *not*
constructive; it merely 'proves' that there exists a fixed point without actually
providing a means to determine (construct) it explicitly[4].

There are now numerous fixed-point theorems and, while they may seem only
of mathematical interest, they have important applications in many branches of
Science, such as in Economics. Indeed, a fixed-point theorem was at the heart of
the game theoretic work of American mathematician John Nash, which earned
him the 1994 Nobel Prize in Economics. The 'Nash equilibrium' point is a fixed
point of a particular continuous function[5], and to prove it, one can use the Brouwer
fixed-point theorem, although Nash originally used an alternative fixed-point theorem attributed to Shizuo Kakutani.

A fine biographical film about John Nash, *A Beautiful Mind*, starring Russell
Crowe and Jennifer Connelly and directed by Ron Howard, came out in 2001.
Like John Nash himself, the film won numerous awards.

---

[4]Thanks to Douglas Bridges for pointing out the irony of Brouwer's non-constructive proof.
[5]A *function* in Mathematics is a relationship between the elements of two sets (discussed in the
subsequent section of the text) that assigns to each element of the first set (the domain) a unique
element of the second set (the range). An example of a function would be that given by the
expression: $y = x^2$, which assigns, to each element x in the domain of real numbers, the unique
real number y, determined by the relationship: $y = x^2$, where y is again a real number. The
domain often depends on context but is generally taken to be as large as possible.

ALL SET

Another issue that arose in the early 20th century was the discovery of some cracks in the very structure of the mathematical edifice of its day. This had to do with the subject of *sets*, which are simply collections of distinct objects, with the objects themselves being the *elements (members)* of the set. For example, the set of letters of the English alphabet can be written as:

$$\{a, b, c, d, e, f, g, h, i, j, k, l, m, n, o, p, q, r, s, t, u, v, w, x, y, z\},$$

and consists of 26 elements. Note also the conventional curly bracket notation, $\{\bullet\}$, for denoting a set. For convenient further reference, we can give the set a name, usually by a capital letter. So, one can write,

$$A = \{a, b, c, d, e, f, g, h, i, j, k, l, m, n, o, p, q, r, s, t, u, v, w, x, y, z\},$$

but the specific letter designation is somewhat arbitrary.

Furthermore, it would seem that we can create sets of almost anything we can think of, such as specific sets of whole numbers like $\{1,2,3\}$, the set of all U.S. States, the set of birthdays of the author (a large set indeed), or even a set which has no elements at all, known as the *empty set*[6].

The theory of sets is an important branch of Mathematics and much of its development stems from the seminal work of the German mathematician Georg Cantor (1845−1918). The above examples of sets are all *finite* sets; that is, one can count up the finite number of elements in any of the sets. Working with finite sets is somewhat mundane and they can even be taught in Primary School.

Cantor, on the other hand, was particularly interested in the exotic realm of *infinite* sets, and in dealing with them we are forced to leave everyday common sense behind. It took a genius like Cantor to figure how to proceed in this infinite realm, as will be seen Chapter 2, although he suffered for his efforts.

Now let us consider one of the bumps in the road of the Mathematical Universe involving sets, namely the Russell Paradox (1901) formulated by the English philosopher Bertrand Russell. He questioned the status of the set,

$S = \{$all sets that are not members of themselves$\}$.

This sounds like a reasonable set to consider. Or is it?

Let us now consider the following two statements:

(i) S is a member of itself.
(ii) S is not a member of itself.

---

[6]The set of observed aliens from outer space would be an example of an empty set, although some who claim to have been abducted by aliens might argue otherwise.

Suppose that *S is* a member of itself, which is statement (i). Then according to the definition of *S*, it is one of those sets that *is not* a member of itself, which is statement (ii). Okay, then suppose that *S is not* a member of itself (statement (ii)). By the very definition of the set *S*, it *is* a member of itself (statement (i)). Therefore, assuming the truth of either (i) or its negation (ii) we arrive at a contradiction. Maybe Brouwer had a point after all.

The preceding dilemma arises from allowing the existence of sets such as *S* in the first place. This has to be more carefully managed, and has been in the current axioms of Set Theory developed by Ernst Zermelo and Abraham Fraenkel in the early 20th century. This will be discussed in Chapter 2, since the most interesting sets are infinite.

Other variations on the Russell Paradox abound:

> *The barber of Seville shaves all and only those men who do not shave themselves.*

If the barber does not shave himself, then according to the statement, he does shave himself. But if the barber does shave himself, then the statement says that he does not shave himself. A way out of this dilemma is to say that no such barber exists.

All this serves to remind us that when we wish to make meaningful statements about the world, we need to proceed with caution. Even the ideal world of Mathematics has some pitfalls to be wary of.

Figure 1.2: The central figures of the famous painting, *Socrates in Athens speaking with Plato,* painted by Raphael. It resides in the Vatican. (Image in public domain.)

WHERE IS MATHEMATICS?

God gave us the integers, all else is the work of man...
German mathematician Leopold Kronecker (1823−1891)

We have had an inkling of mathematical reasoning, but what of Mathematics itself? What exactly is it? Some investigators, like neuropsychologist Brian Butterworth, have argued on evolutionary grounds that the human brain is hard-wired for numeracy (see the Bibliography for his very interesting account).

The author would argue that the human genome – the full set of genes that make us what we are – contains instructions for building specialized circuits of the brain, which he calls the Number Module. The job of the Number Module is to categorize the world in terms of numerosities – the number of things in a collection…

Our Mathematical Brain, then, contains these two elements: a Number Module and our ability to use the mathematical tools supplied by our culture.

Those tools would include counting on one's fingers for starters, as well as an abacus, hand calculator, and all the other calculating devices that history has provided us with. Of course, trade and commerce made numeracy an essential ingredient and hastened the development of arithmetic.

For many people, Mathematics is just some form of glorified arithmetic. Or if they studied some algebra or geometry in high school, people often confess that they were never very good at it. Humorist Fran Lebowitz nicely sums up a prevailing view when she states that, "In real life, I assure you, there is no such thing as algebra[7]." This is not only a good joke, but it contains an element of truth in it, in so far as it makes the point that algebra is actually a mathematical abstraction. You will not see it anywhere on the streets of New York, Fran.

On the other hand, the concept of number is also an abstraction, so we are forced to conclude that we are going to have to deal with abstractions if we are to talk about Mathematics at all.

Indeed, besides the abstract notion of number, Mathematics inhabits a world of perfect circles, straight lines, triangles with exactly 180 degrees and so forth. Yet this is only an idealization and the world we live in can never contain a perfect circle or perfectly straight line. All lines will have some variations from absolute perfection, although these may be exceedingly small. They will also have some

---

[7] Because many people who have encountered algebra in school have found it so arcane, it has attracted a considerable body of humor:

"*When you are dissatisfied and would like to go back to youth, think of algebra*" … Will Rogers

"*What is algebra exactly; is it those three-cornered things?*" James M. Barrie

"*As long as algebra is taught in school, there will be prayer in school*" … Cokie Roberts

width, whereas a line in Mathematics has none. Numbers also go on forever, yet nothing in our experience seems to go on forever, except perhaps boring lectures about Mathematics. Even the Universe itself is expected to come to an end – more on the end of the world later.

Yet Mathematics is the most powerful tool we have that allows us to describe the world. Why should this even be so? In a certain sense, subscribed to by the author, Mathematics lies outside this world, in an 'ideal world' postulated by Plato (Fig. 1.2). Thus, mathematical concepts seem to occupy some transcendent realm (a view called *Platonism*) existing outside time and space.

How this transcendent ideal world interacts with the real world is the subject of this book and when we get to Quantum Mechanics, the real world will also take on the appearance of something quite other worldly as well.

Let us take one historical example that most people are familiar with: the 2,500-year-old great theorem of Pythagoras, which tells us that in any right triangle, the square of the length of the hypotenuse is equal to the sum of the squares of the lengths of the other two sides (Fig. 1.3).

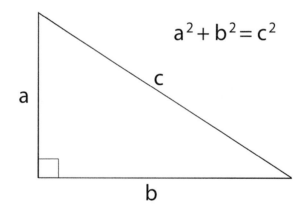

$$a^2 + b^2 = c^2$$

Figure 1.3: The famous theorem is attributed to Pythagoras (ca. 570–495 B.C.) although the evidence for a proof by him is sketchy and the result had been known since earlier times. (Illustration courtesy of Katy Metcalf.)

But again, this is only an idealization that is valid in the Mathematical Universe, and any actual right triangle that we can produce will have some minor discrepancy from what the theorem states. But the more accurately we draw a right triangle, the more precise the result will become.

On the other hand, this ancient result named after Pythagoras, perhaps initially drawn in the sand, has had countless applications to our understanding of the real world. It even makes a crucial appearance in Einstein's Theory of Relativity regarding how time in motion slows down, which is at the heart of the Global Positioning System (GPS)[8].

There is also a converse of the theorem; that is, for any triangle having sides of length a, b, c, that satisfy the relationship

$$c^2 = a^2 + b^2,$$

then the angle formed by the sides of lengths a and b is a right angle (= 90°). Proofs of both theorems appear in Euclid's *Elements*, discussed later in this chapter.

Mathematical theorems, like that of Pythagoras, are seemingly 'discovered'. A sprawling labyrinth of mathematical discovery arose from the concept of the 'imaginary number' which represents the square root of −1, and which is the subject of Chapter 3. It is a beautiful body of work, known as *complex analysis*, that is the basis for a lot of 'real' Science from Quantum Mechanics, to electrical circuitry, the Theory of Relativity, or the design of airplane wings... Yet the mathematics of complex analysis is not something concrete. It exists outside our physical world, dare we say in a universe of its own.

Roger Penrose, certainly one of the greatest living scientific minds on the planet, says, "I have been arguing that such 'God-given' mathematical ideas should have some kind of timeless existence, independent of our earthly selves." The author will not venture a guess as to whether this is a theistic statement or not.

The eccentric, itinerant (and atheist) Hungarian mathematical genius, Paul Erdös, often mentioned a book in which God had recorded all the most elegant and beautiful mathematical proofs. This sentiment reflects the notion, felt by many mathematicians, that Mathematics is fundamental to the very nature of the Universe.

Similarly, Godfrey Harold (G.H.) Hardy, one of the 20th century's finest mathematicians, held a similar view: "I believe that mathematical reality lies outside us, that our function is to discover or observe it, and that the theorems which we prove, and which we describe grandiloquently as our 'creations', are simply our notes of our observations. This view has been held, in one form or another, by many philosophers of high reputation from Plato onwards...[9]"

On the other hand, many have argued against the Platonist perspective and take the view that Mathematics is a game played by mathematicians based on a set of logical rules and a set of axioms. No set of axioms is to be preferred over any other. This is a view more readily maintained by a non-mathematician and is held by a number of philosophers.

---

[8] See the section Time in Motion of Chapter 7, as well as Appendix XV where it is explicitly utilized.

[9] *A Mathematician's Apology*, G.H. Hardy, 1940, p. 123.

Relevant to the question of whether Mathematics is merely a mental construct of mankind, or whether it already exists in some netherworld, is the enduring mystery of *why* Mathematics is so remarkably useful in the scientific exploration of the real world. Nobel Laureate in Physics, Eugene Wigner, summed up the situation thusly in his 1960 article entitled: *The Unreasonable Effectiveness of Mathematics in the Natural Sciences*[10]: "The enormous usefulness of mathematics in the natural sciences is something bordering on the mysterious, and there is no rational explanation for it... it is an article of faith."

The article stirred up a hornet's nest of controversy among scientists and philosophers, again with many of those agreeing being scientists and many of those disagreeing being philosophers. The reader is invited to make up their own mind by the end of this book.

The situation is nicely summed up by cosmologist Brian Greene, who is himself in two minds about the Platonist vs. non-Platonist views of what is Mathematics:

> "I could imagine an alien encounter during which, in response to learning of our scientific theories, the aliens remark, 'Oh, math. Yeah, we tried that for a while. At first it seemed promising, but ultimately it was a dead end. Here, let us show you how it really works'. But to continue with my own vacillation, I don't know how the aliens would actually finish the sentence, and with a broad enough definition of mathematics (e.g., logical deductions following from a set of assumptions), I'm not even sure what kind of answers *wouldn't* amount to math[11]."

FINE TUNING

With the aid of Mathematics, one remarkable discovery has been that our Universe is exceedingly finely tuned to our being here in the first place. This seems to present another mystery as to why that should be so. If any of the numerous physical constants found in Nature were even slightly different from what they are, our Universe would have evolved completely differently, or not at all, and we would likely not be here to discuss it. For example, according to theoretical physicist Max Tegmark of the Massachusetts Institute of Technology, "If protons were 0.2% heavier, they'd decay into neutrons unable to hold on to electrons, so there would be no atoms[12]."

---

[10] *Comm. Pure and Appl. Math.*, 13, 1–14, 1960. This was the Richard Courant lecture in mathematical sciences delivered at New York University, May 11, 1959.

[11] Brian Greene, *The Hidden Reality*, p.297.

[12] M. Tegmark, *Our Mathematical Universe*, p. 142.

In Tegmark's view, there is another perspective regarding the nature of Mathematics, one that could provide a resolution to the mystery of why it is so useful in describing the Universe. For Tegmark, the universe of Mathematics *is* the Universe. According to his Mathematical Universe Hypothesis (MUH), our Universe is itself a mathematical structure and that is why Mathematics is so successful in explaining it. In an attempt to avoid the difficulties of infinite regress within this mathematical world, Tegmark claims[13]:

> "[A]t the bottom level, reality is a mathematical structure, so its parts have no intrinsic properties at all! In other words, The Mathematical Universe Hypothesis implies that we live in a *relational reality*, in the sense that the properties of the world around us stem not from properties of its ultimate building blocks, but from the relations between these building blocks. The external physical reality is therefore more than the sum of its parts, in the sense that it can have many interesting properties while its parts have no intrinsic properties at all."

The notion of the *relational reality* that Tegmark is alluding to is a common feature of various structures found in Mathematics. One of these, known as a *group*, is discussed in Chapter 7. This is simply a set whose elements obey certain *relational* properties to one another. The elements themselves do not have any intrinsic properties − they could even be letters of the alphabet − but what makes the set a group is how all the elements relate to one another.

Tegmark also envisions the possibility of four different categories of multiverses (other Universes besides our own), some with the same laws of Physics, some with different laws involving different constants of Nature.

Distinguished British cosmologist Martin Rees has boiled down the critical parameters of Nature to six in number (see Bibliography), two of which, the density of matter in the Universe (such as gram/cm^3) and the cosmological constant Λ, representing an intrinsic property of empty space[14], are much discussed in the text. Taking one example from Rees:

> "The cosmos is so vast because there is one crucially important huge number N in nature, equal to 1,000,000,000,000,000,000,000,000,000,000,000,000,000. This number measures the strength of the electrical forces that hold atoms together, divided by the force of gravity between them. If N had a few less zeros, only a short-lived miniature universe could exist: no creatures could grow larger than insects, and there would be no time for biological evolution."

---

[13] Ibid, p.267.

[14] 'Empty space' contains a considerable amount of energy, which plays an important role in the future expansion of the Universe. See the discussion following Hotlink eq. (30) in Chapter 7.

Attempting to build a Universe with either two or four spatial dimensions is also a non-starter, as both have problems with gravity. Many such arguments have been advanced by others to indicate the exquisite 'fine-tuning' of the essential parameters of our Universe, so that it could evolve over sufficient time and space in order to lead to stars, galaxies, planets, and ultimately life. Here we are now to exalt in its creation. Indeed, some have viewed this fine tuning as evidence of a Creator who had their hand on the dials, such as in the book, *God's Undertaker – Has Science Buried God?* by Oxford mathematician John C. Lennox (see Bibliography).

The view that multiverses provide a solution to the quandary of why the physical constants of the Universe happen to be so perfectly attuned for our existence has become very popular recently, as espoused in the excellent books by Brian Greene (see Bibliography). Some members of the multiverse will die out quickly, as envisaged by Rees above, while some will expand too rapidly for stars and galaxies to form, and some will consist of *absolutely nothing*. This latter Universe neatly answers the question, 'Why is there something rather than nothing?' There is a 'nothing Universe', we are just not in it.

It should also be noted that some scientists and philosophers reject the notion of fine-tuning altogether. Germane to this whole discussion is the '*Anthropic Principle*' enunciated by Australian physicist Brandon Carter in 1973, although the notion has occurred previously to others. There are several versions of the principle, but essentially it comes down to: *The very fact of our existence means that the fundamental constants of our Universe must be fine-tuned so as to allow for human existence.* In other words, if the constants were not just so, we would not be here to discuss the matter. For a very comprehensive discussion of this fascinating topic, see the work by Barrow and Tipler in the Bibliography.

Of course, even this proposal is not without controversy, and indeed, the whole issue of the fine tuning of the physical parameters that allow for our existence is a deep matter involving Physics, Cosmology, Philosophy and Religion. A fine account of the matter is presented in Paul Davies' book, *The Goldilocks Enigma: Why Is the Universe Just Right for Life?* (see Bibliography).

The term 'Goldilocks' alludes to the story of *Goldilocks and the Three Bears*, a British nursery tale cherished by millions of children since the 19th century. The young girl star of the story, Goldilocks, enters the household of three bears who happen to be out in the forest waiting for their porridge to cool. She in turn tests their porridge (too hot/too cold/just right), seats (too big/also too big/just right), and beds (too hard/too soft/just right). Of course, there is more to the story …

The fairy tale has been adopted by astronomers to describe when something happens to be 'just right' for life. Like our Earth happens to be at just the right distance from our star the Sun. If it was closer like the planet Venus, we would fry.

Further out like Mars and we would freeze. So, our planet happens to be located in what is called the 'habitable zone' which is just right for us (and for Goldilocks, as well as the three bears).

Thus, the bottom line for us Earthlings is that we live in a Goldilocks Universe, at a Goldilocks period in its evolution, on a Goldilocks planet, at a Goldilocks distance from its nearest star. Fortunately, we do not have to take sides with any of the preceding debates because we are only considering the Mathematics applicable to our Universe as we know it. But we will mention a random element to our existence which should not be overlooked.

One striking piece of dumb luck came our way that has nothing to do with the fine tuning of universal parameters. During the early formation of the Earth some 4.56 billion years ago, there were a large number of planetesimals (small rocky planet-like bodies) flying about the Solar System in various orbits. One in particular, now named *Theia* and roughly the size of Mars, struck the Earth in a head-on collision, knocking it off its vertical axis (by 23.5 degrees), with the debris that was blasted into space eventually coalescing to form the Moon (Fig. 1.4)[15].

As a consequence, the tilt of the Earth's axis now gives us the seasons and the Moon gives us the tides, as well as stabilizing the Earth's rotation.

Even the formation of life on Earth has been affected by this random collision event. A 2019 study has found that it is likely much of the Earth's carbon, nitrogen and other volatile elements essential for life came from the Earth-Theia collision[16]. The study's lead author, Damanveer S. Grewal, claims "This study suggests that a rocky, Earth-like planet gets more chances to acquire life-essential elements if it forms and grows from giant impacts with planets that have sampled different building blocks, perhaps from different parts of a protoplanetary disk[17]."

Moreover, the ingredients that make up life itself may have also been a consequence of the Earth-Theia collision. According to the molecular biologist Richard Lathe, "[Thc] Odds of nucleic acids [DNA and RNA] forming on Earth without the lunar tides would be much lower[18]."

As a consequence, the publication of this book may have been delayed by millions of years but for the Earth being blindsided in a random collision from space.

---

[15] This is currently the most widely accepted theory for the formation of the Moon and the tilt of the Earth's axis. However, the theory is still incomplete regarding some aspects. The name derives from Greek mythology and Theia's role as mother of the Moon goddess.

[16] D. Grewal, *et al.*, Delivery of carbon, nitrogen, and sulfur to the silicate Earth by a giant impact, *Science Advances,* 5, no. 1, Jan 23, 12 pp., 2019.

[17] The proto-planetary disk was the swirling mass of gas and dust around our proto-Sun, out of which the planets formed, including our own.

[18] R. Lathe, Without the Moon, Would There Be Life on Earth?, *Sci. Amer.*, 2009.

Figure 1.4: Random events have played a significant role in our existence on Earth. (Image courtesy of NASA/JPL-Caltech.)

Even more good fortune came our way 66 million years ago, when a freak encounter with a 10 km rock from space led to the extinction of the dinosaurs and paved the way for human existence. Again, it is questionable whether you would be reading this book were it not for this freak event. It is difficult enough getting along with other members of our own species, so it is impossible to envisage how dinosaurs and humans could ever manage to co-exist. Indeed, a good many people are killed by crocodiles each year (and several by alligators) and they are not even true dinosaurs, although they are closely related. Thus, we should make the most of our stunning random good luck.

A BLAST FROM THE PAST: EUCLID'S GEOMETRY

The postulates of Mathematics were not on the stone tablets that Moses brought down from Mt. Sinai...
American mathematician Richard Hamming

Certainly, one of the greatest and most influential treatises on mathematical thought was the *Elements*, by the Greek mathematician Euclid of Alexandria. The works comprise 13 volumes and although written around 300 B.C., their contents have managed to curse the lives of many a student for the next 23 centuries (Fig. 1.5). But more profoundly, the *Elements* established a formal framework in which the whole of geometry was to be conducted, namely the laying down of a few transparently obvious axioms and the deduction by logical principles of the consequences from these axioms, which become known as theorems (or propositions).

Figure 1.5:   Euclid demonstrating a theorem to his students. From the same painting by Rafael as in Fig. 1.2. (Image in public domain.)

In the words of Paul Erdös again, "A mathematician is a machine for turning coffee into theorems." (Although he may not have been the first to utter this, the author heard Erdös say it). Mathematics, the stuff that university mathematicians do, is all about proving new theorems set in an axiomatic framework. Most people are surprised to learn that anything new can actually be discovered in Mathematics, but as it turns out, new discoveries are being made on a daily basis, essentially by turning coffee (and axioms) into new theorems.

One of the main differences between Mathematics and Science is that in Mathematics, results are 'proved', whereas in the physical sciences this is never the case. In Mathematics, there are various assumptions that are held to be self-evident. For example, in Euclid's geometry, there were five axioms or postulates, which are for the most part obvious to anyone who has ever thought about them. Yet from these five axioms of Euclid flow, by pure deductive reasoning, the entire fields of plane (two-dimensional) and solid (three-dimensional) geometry. It should be noted, however, that a fair amount of Euclid's work was based upon the earlier work of other Greek scholars.

What is not so widely appreciated, by students who encounter Euclid's geometry for the first time, is that the very idea of creating an axiomatic system of self-evident postulates from which to derive all of geometry is where the genius lies.

The universe of plane geometry that many of us encountered in school is an idealized one of infinite flat planes, perfectly straight lines that have no width, points that have no size, only a location, perfect circles, and so forth. So, in order to construct the entire edifice of the perfect universe of Euclidean Plane Geometry, all we need are five elementary postulates. We give them here just to refresh our memory, not that we will do anything much with them, except for the very last one:

1. *A straight-line segment can be drawn connecting any two points.*
2. *Any straight-line segment can be extended indefinitely in a straight line.*
3. *Given any straight-line segment, a circle can be drawn having the segment as radius and one endpoint as its center.*
4. *All right angles are congruent[19].*
5. *If two lines are drawn which intersect a third in such a way that the sum of the inner angles on one side is less than two right angles, then the two lines eventually intersect each other on that side if extended far enough.* This postulate is known as the *Parallel Postulate* (see Fig. 1.6a).

Because of its very specific nature, people historically thought that Euclid's 5th Postulate could be deduced as a consequence from the other postulates, and several mathematicians historically gave (ultimately) erroneous proofs of the 5th

---

[19] A right angle is formed by two lines meeting at 90 degrees, and congruent means that any two right angles are exactly the same except for position and orientation in the plane. In other words, a right angle is a right angle, is a right angle…

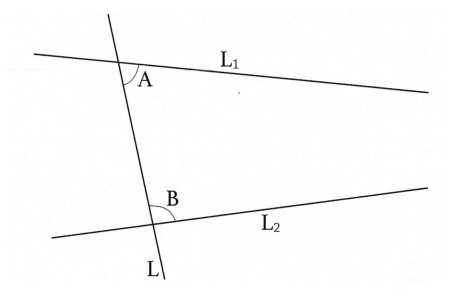

Figure 1.6a: Euclid's 5th Postulate, whereby the line L makes two angles A and B that are less than two right angles. Hence, the lines L_1 and L_2 will eventually intersect. The word 'eventually' is a little ambiguous in this situation. (Illustration courtesy of Katy Metcalf.)

Postulate. It turns out that this is not the case; that is, it is *independent* of the other postulates and so cannot be proved from them[20]. The 'eventual' meeting of the two lines in question is not a very satisfactory state of affairs. What if one has to traverse much of the entire Universe to find the intersection point? Fortunately, there is an entirely equivalent formulation to the 5th Postulate in the context of the other four axioms, the well-known *Playfair Axiom*, which is more transparent (Fig. 1.6b)[21]:

> *In a plane, given a line and a point not on it, at most one line parallel to the given line can be drawn through the point.*

Figure 1.6b: Given a straight line L and a point p not on L, there is at most one line parallel to L through the point p.

---

[20] This has not stopped many amateur mathematicians from providing 'proofs' of the 5th Postulate into the present day.

[21] Due to Scottish mathematician John Playfair (1748−1819).

In fairness to Euclid, it should be noted that Playfair's Axiom has a slight deficiency, in that it does not explain how to construct the straight line through the point and parallel to the given line. It is only asserted that there is such a line.

The point of Euclid's postulates is that there is a notion of 'mathematical truth', which applies to this idealized world of Mathematics but does not hold in other physical sciences. If one follows the basic principles of mathematical deduction, of deriving one step from another, mathematical truths (theorems) are obtained and become additional parts of the pantheon of this idealized world. They become eternal unshakable truths. That is one of the great attractions for mathematicians to their subject. It possesses this wonderful idealized notion of truth. This is not to say that there are no paradoxes in Mathematics, as we have already encountered in Russell's Paradox among others.

Of course, in a physical Science, such as Chemistry, Biology, or Physics, the situation is quite different and often one theory will replace another as more data and understanding are obtained.

Another important consideration is that a given length in Mathematics, say the line between the points 0 and 1, can be indefinitely divided. One can take half the distance from 0 to 1, one-third the distance, one-fourth, one-millionth, one-trillionth, and so on. This is not necessarily so in real life. Whether or not space itself can be divided infinitely is an important question and is discussed further in Chapter 2.

In spite of the distinct nature between the idealized mathematical world and the imperfect real world, the great physicist Richard Feynman, echoing Wigner, summed it up by saying that it is "quite amazing that it is possible to predict what will happen by mathematics, which is simply following rules which really have nothing to do with the original thing." Examples of this predictive power of little squiggles on paper will be presented later in the text.

TAKING THE FIFTH FURTHER

The 5th Postulate of Euclid is somewhat different from the others, in that it says that two lines will *eventually* cross under the right circumstances. In the Playfair version, only one line can be drawn through a point that will be parallel to a given line, and that any other will not be parallel and hence will cross the given line at some stage. This uneasy feeling about the nature of the 5th Postulate has led some researchers to scrap it and replace it with something else. The result would then be a new kind of geometry.

It should be clear that Euclidean geometry is not going to be applicable on a saddle-shaped surface or on a sphere, as per Fig. 1.7. In Euclidean geometry, *the sum of the angles of any triangle is equal to two right angles; that is,*

$180°(= \pi$ *radians*$)^{22}$. This statement, known as the *Triangle Postulate*, happens to be equivalent to the statement of Euclid's Parallel Postulate.

However, on the surface of a saddle the angles of a (*hyperbolic*) triangle sum to less than $180°$, and on a sphere the sum of the angles of a (*spherical*) triangle formed by great circles is greater than $180°$ (Fig. 1.7). Therefore, as the Triangle Postulate does not hold, the Parallel Postulate also does not hold on either of these two surfaces.

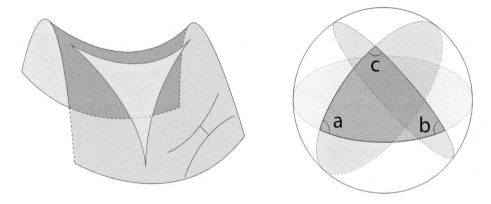

Figure 1.7:   (L) The geometric figure of a hyperbolic paraboloid (commonly known as a saddle) is an example of a surface said to have 'negative curvature'; that is, the angles of a (hyperbolic) triangle on the saddle sum to less than $180°$. A flat plane has 'zero curvature' where the angles of any triangle add up to exactly $180°$. The sphere (R) has 'positive curvature', with the angles of a (spherical) triangle adding up to more than $180°$. Thus, the Parallel Postulate does not hold on either the surface of a saddle or the surface of a sphere. See also Fig. 6.12 in Chapter 6. (Illustration courtesy of Katy Metcalf.)

It turns out that on the surface of a sphere, where a 'straight line' is a portion of a great circle[23] as in Fig. 1.7 (R):

*Given a line and a point not on a line L, there is **no** straight line that passes through the point that does not intersect L,*

which the great German mathematician Bernhard Riemann (1826–1866) took to replace Euclid's 5th Postulate in his development of Riemannian Geometry.

---

[22] Radians are generally used by mathematicians instead of degrees. To convert degrees to radians, you multiply degrees times $\pi/180$. So for example, $90° = \pi/2$ radians, $45° = \pi/4$ radians, $30° = \pi/6$ radians, and so forth. One radian equals approximately 57.3 degrees and is the angle formed in a sector of a circle when the arc on the circle's circumference has the same length as the radius of the circle. It is not important to know this fact.

[23] A great circle is the largest circle that can be drawn on the surface of the sphere, as in Fig. 1.7 (R).

On the other hand, on the surface of a saddle:

*Given a line and a point not on a line L, there is **more than one** straight line that passes through the point that does not intersect L,*

which one can take to replace Euclid's 5th Postulate to form the basis of another new geometry very similar, yet fundamentally different from Euclid's. This approach was taken almost simultaneously by two mathematicians, the Hungarian, János Bolyai (1802–1860)[24] and the Russian, Nikolai Lobachevsky (1792–1856), although the latter published first[25]. This new geometry, known as Hyperbolic Geometry, is pertinent to the Theory of Relativity and has rather intriguing properties, as we are about to find out.

Of course, the mathematical giant, Carl Friedrich Gauss (1777–1855), beat both Bolyai and Lobachevsky to it but did not publish his results, fearing the ensuing controversy.

Instead of thinking about Hyperbolic Geometry in the plane, which is hard to grasp due to its infinite extent, it is actually possible to visualize the whole geometric set up in the interior of a circle. But you will have never seen a circle quite like this before.

This exotic region where Hyperbolic Geometry plays out is known as the two-dimensional 'Poincaré disk model', named for French mathematician and theoretical physicist Henri Poincaré (1854–1912), one of the giants of modern Science. It is simply a flat round disk that is the interior of a circle of radius = 1 which is centered at the origin and does not include the boundary. The entire hyperbolic geometric universe resides within this disk, as in Fig. 1.8. Its handy size makes it convenient to visualize all the action taking place inside. It is related to a 'hyperboloid' in three-dimensional space, which is a bowl-shaped figure discussed in detail in Appendix II.

In geometry, you need a notion of straight lines. In this model, 'straight lines' are arcs of circles (e.g. L, L_1, L_2, L_3 in Fig. 1.8) that are perpendicular to the boundary but theoretically never reach the boundary of the disk, which is at 'infinity'. So, in some sense we can visualize the edge of our space at infinity which coincides with the points of the circle of radius =1.

In this model, Euclid's 5th Postulate does not hold; that is, there is more than one straight line through a given point that is parallel to a given line. Indeed, there are infinitely many, as in Fig. 1.8.

---

[24] Interestingly, the father of János, mathematician Farkas Bolyai, was one of those who gave several false proofs of the Parallel Postulate, advising his son against working on the problem: "For God's sake, please give it up. Fear it no less than the sensual passion, because it, too, may take up all your time and deprive you of your health, peace of mind and happiness in life." Sometimes, it is best if sons do not listen to the advice given by their fathers.

[25] Lobachevsky himself demonstrated a false proof of the Parallel Postulate in a geometry course in 1815.

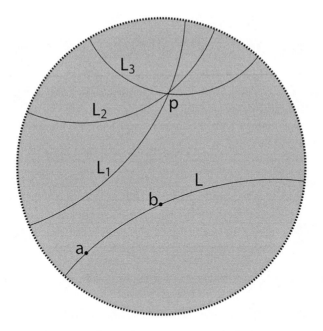

Figure 1.8: An entirely new form of geometry (Hyperbolic Geometry) can be created that takes place inside the Poincaré disk model, in which 'straight lines' are arcs of ordinary circles that are perpendicular to the boundary. Euclid's 5th Postulate is violated, as there are infinitely many 'straight lines' passing through a given point p and not intersecting a given line L as displayed in this image. (Illustration courtesy of Katy Metcalf.)

As the reader can see, the Poincaré disk model represents no ordinary geometry. However, it has the nice feature that it all evolves inside the circle of radius $r = 1$. Some of it will look very familiar in just a moment.

A 'straight line' connecting two points, such as a and b as in Fig. 1.8 in this disk model, is an arc of an ordinary circle (called a *geodesic*) between the two points, with the extensions of the arc perpendicular to the boundary. This is essentially just what we all learned in school – namely that the shortest distance between two points is a straight line (segment) – but here the 'straight line' is a hyperbolic circular one as in Fig. 1.8[26].

The Euclidean distance from the center of the disk at the origin 0 to some point on the number line, say $r = 0.5$, is just 0.5. In general, the ordinary Euclidean distance from the origin to any value r on the number line is just r. We all know this already.

---

[26] Here, we are blurring the distinction between a line segment connecting two points and the infinite extension of the line, so as not to be pedantic.

On the other hand, the 'hyperbolic distance' to the point 0.5 happens to be 1.1^{27}, while the hyperbolic distance to the point 0.99 is 5.3. Indeed, as we approach ever closer to the boundary of the circle at $r = 1$, the hyperbolic distance to that point becomes larger and larger, towards becoming infinite, albeit rather slowly. The reason of course is that we are moving closer and closer to our 'infinity' and we have an infinite space represented inside a finite one.

The whole of geometry can be accommodated in the Poincaré disk and, indeed, it is a Mathematical Universe in its own right. For example, the sum of the angles, say a, b, c, of a *hyperbolic triangle* are strictly less than $180° = \pi$ radians (like on the saddle in Fig. 1.7.) The *hyperbolic area* of such a triangle is given entirely by the size of the three angles that form the triangle; that is, by the (Gauss-Bonnet) formula,

$$A_h = \pi - a - b - c.$$

From this formula, one can see that all hyperbolic triangles having the same angles (a, b, c) have the same hyperbolic area[28], and that triangles whose three angles are zero radians ($= 0°$), have the largest hyperbolic area $= \pi$. Remember, the sides of a hyperbolic triangle are curved and so that is how such a triangle can have angles with zero degrees.

In fact, the exotic nature of hyperbolic disk geometry is a feature in numerous works by Dutch artist M.C. Escher (as in Fig. 1.9 (R)) and so we have seen this all before.

Let us consider one of the many tessellations of the Poincaré disk, as in Fig. 1.9 (L). The three vertices of each one of the (hyperbolic) triangles meet with other hyperbolic triangles, and depending on where one starts counting, there are 4, 8, and 12 (hyperbolic) triangles meeting respectively at each vertex. Where we find *four* triangles meeting, the corresponding angle of each triangle at that common vertex is 90°; where *eight* triangles meet, each angle there is 45°; where *twelve* triangles meet, each angle at the vertex is 30°. Note that the sum of the angles of each hyperbolic triangle is therefore 165° which is less than the sum of 180° for the interior angles of any Euclidean triangle.

These three interior angles remain the same for every hyperbolic triangle of the tessellation, even for those that are apparently getting 'smaller and smaller', in the Euclidean sense, as one nears the boundary circle at infinity. But actually, the triangles are not getting smaller and smaller in the hyperbolic area sense. Since

---

[27] The hyperbolic distance from the origin to a point r on the number line happens to be given by the formula: $H_{dist} = \ln\dfrac{1+r}{1-r}$, where *ln* is the natural logarithm. Putting in $r = 0.5$ gives the hyperbolic distance as 1.1. The reader is invited to try $r = 0.99$. This hyperbolic distance will become infinite as r approaches the value 1, but glacially slowly due to the painfully slow growth of the logarithm function.

[28] In the language of high school geometry, similar triangles are congruent.

every triangle has the same three interior angles, 90°, 45°, and 30°, they all have the same hyperbolic area according to the preceding Gauss-Bonnet formula. We leave to the reader the arithmetic to compute the area.

Moreover, since the whole disk is composed of infinitely many such triangles, the hyperbolic area of the disk is likewise infinite, even though in the Euclidean sense the area of the disk is: $A = \pi r^2 = \pi$, since our radius is $r = 1$.

Figure 1.9: A tessellation of the Poincaré disk with the number of hyperbolic triangles meeting at the vertices of each triangle being 4, 8, and 12 (count them for yourself). Thus, the corresponding angles of each hyperbolic triangle are: 90°, 45°, and 30° respectively, and therefore they all have the same hyperbolic area. Various tessellations of the Poincaré disk can be created with different hyperbolic polygons. (R) The hyperbolic disk inspired Dutch artist M.C. Escher to create a series of four woodcut images attempting to capture the essence of infinity. The one here is Circle Limit IV Angels and Devils (1960). (Images in public domain.)

Why have we bothered with hyperbolic space, besides paying a tribute to artist M.C. Escher? One important reason is that it features strongly in the geometry of what is called 'spacetime' (see Appendix II, where it appears in Minkowski Spacetime), which joins three-dimensional space with a time dimension via a particular geometry, such as Hyperbolic Geometry.

Indeed, there is one particular model of spacetime, known exotically as the 'anti-de Sitter model[29]' that consists of a stack of Poincaré disks much like a stack of casino chips. Actual casino chips have a thickness of a few millimeters, but our

---

[29] This model arises from a solution of Einstein's field equations given in Chapter 7, where Dutch mathematical physicist Willem de Sitter re-enters the scene.

Poincaré disks are two-dimensional mathematical disks and so have no thickness. Nevertheless, let us mentally conflate the two so that in the anti-de-Sitter model, each casino chip (Poincaré disk) in the stack represents the state of the Universe at a particular instant of time, as in Fig. 1.10. The amalgamation of the stack of casino chips (Poincaré disks) results in a solid cylinder with time, representing the height of the cylinder, being the third dimension. This stack of disks then becomes, in some sense, a model of the evolution of the Universe over time.

If you are wondering how stack of disks with no thickness can actually form a cylinder, then you have asked a very reasonable question. To answer it, let us consider an even simpler situation, namely all the points on the number line between the values of 0 and 1. This set of points is infinite, yet each point has zero dimension; in particular, a single point has no length. But when we take the infinite set of *all* the points between 0 and 1, we have a line of length =1. This line segment is a type of infinite set known as a *continuum*, which will be discussed more fully in the next chapter.

In the case of our infinite stack of disks, the vertical time axis is a continuum of infinitely many instances of time and, in this case, the disks corresponding to each instant will form a solid cylinder.

Of course, this anti-de-Sitter model has only two spatial dimensions and the actual Universe that we live in has three, but the Hyperbolic Geometry model does have a higher dimensional analogue. Indeed, the boundary of the Poincaré disk, which is a circle 'at infinity' in the two-dimensional case, is replaced by a sphere 'at infinity' in the three-dimensional case, with Hyperbolic Geometry taking place inside the sphere. Adding a further dimension of time, as we did by stacking the Poincaré disks, we again have an anti-de-Sitter space and a more realistic (but harder to visualize) model of spacetime.

One remarkable consequence of modern theoretical Physics is that information from *inside* the (higher dimensional) anti-de Sitter space is encoded on the boundary, in the way that a hologram encodes information in a two-dimensional image about a three-dimensional scene.

Thus, a gravitational theory in the interior anti-de Sitter space can be transferred to the boundary where it is possible the calculations are easier. This idea has profound implications in Quantum Mechanics, which is the Physics of the atomic and sub-atomic universe of particles and forces, to be discussed in a subsequent chapter. Therefore, problems relating to black holes can also be resolved with the *Holographic Principle*[30], as a black hole corresponds merely to a configuration of particles on the boundary without the complicating factor of having to deal with gravity[31].

---

[30] The Holographic Principle was originally proposed by Dutch theoretical physicist Gerard 't Hooft and developed by Leonard Susskind and others.

[31] Juan Maldacena, The Illusion of Gravity, *Sci. Amer.*, Nov. 2005, pp. 57–63.

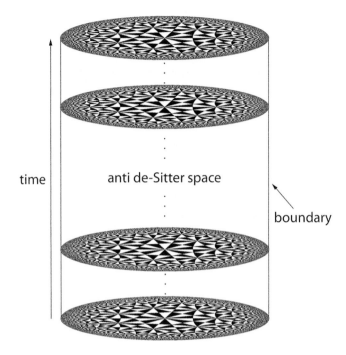

time   anti de-Sitter space

boundary

Figure 1.10: By stacking Poincaré disks, each representing the Universe at a given instant of time, we obtain a cylinder with the evolution over time depicted by its height. The boundary of the anti-de Sitter model can be used to explore what is happening in the interior mathematically according to the 'Holographic Principle', and vice versa. (Illustration courtesy of Katy Metcalf.)

According to theoretical physicist Juan Maldacena of the Institute for Advanced Study, "Amazingly, some new [holographic] theories of physics predict that one of the three dimensions of space could be a kind of an illusion – that in actuality all the particles and fields that make up reality are moving about in a two-dimensional realm like the Flatland of Edwin A. Abbott[32]. Gravity, too, would be part of the illusion: a force that is not present in the two-dimensional world but that materializes along with the emergence of the illusory third dimension."

---

[32] *Flatland*, by E.A. Abbott was published in 1884 and describes a fictional two-dimensional world inhabited by lines (women) and polygons (men) that was meant as social satire of the period.

It should be mentioned, however, that "So far no example of the holographic correspondence has been rigorously proved – the mathematics is too difficult[33]." But according to theoretical physicist Raphael Bousso, "The world doesn't appear to us like a hologram, but in terms of the information needed to describe it, it is one." Indeed, there is some supporting evidence for holographic models of the very early Universe[34].

As well as spacetime, something much more down to Earth, like an array of complex networks such as the internet, also falls under the purview of Hyperbolic Geometry. "[W]e develop here a geometric framework to study the structure and function of complex networks... We begin with the assumption that hyperbolic geometry underlies these networks... Conversely, we show that if a network has some metric structure, and if the network degree distribution is heterogeneous, then the network has an effective hyperbolic geometry underneath[35]."

Gauss, Lobachevsky, and Bolyai had no idea what their strictly theoretical, strange looking geometry would lead to.

PI IN THE SKY

> Apu: *In fact, I can recite pi to 40,000 places. The last digit is one!*
> Homer: *Mmmm, pie.*

Quite likely, everyone who has ever attended school has heard of *pi* or π. This symbolic notation that we still use today has been around since 1737, when it was used by the great mathematician Leonhard Euler (pronounced 'oiler'). The ratio of the circumference of any circle to its diameter has a certain global fame that few Hollywood stars can match. Now, we even celebrate a 'Pi Day' on March 14 (3-14, which also turns out to be the date that Albert Einstein was born in 1879) for all pi enthusiasts[36]. There is also an excellent book, *Life of Pi*, by Yann Martel, although Pi in this instance is the name of the main character.

---

[33] Juan Maldacena, The Illusion of Gravity, *Sci. Amer.*, Nov. 2005, p. 63.

[34] N. Afshordi, *et al.*, From Planck data to Planck era: Observational tests of Holographic Cosmology, *Phys. Rev. Lett.*, 118, 8 pp., 2017.

[35] Dmitri Krioukov *et al.*, Hyperbolic geometry of complex networks, *Physical Rev.* E 82, 2010.

[36] As the British write the date in day/month order, they prefer July 22 (22/7) for Pi Day as 22/7 = 3.14... to two decimal places. As an aside, that also happens to be the birthday of German mathematician and physicist Friedrich Wilhelm Bessel (1784–1846), whom we encounter in footnote 23 of Chapter 2. The Bessel differential equation and subsequent solutions (Bessel functions) play an important role in many branches of Physics.

The intrinsic notion of π comes built into the fabric of our Universe. The length of the circumference of any circle is equal to π times its diameter, no matter how large or small. Of course, these circles are mathematical idealizations, as we know that any physical circle that we measure will have slight imperfections in its roundness and there are also difficulties with making extremely precise measurements of any real object. So π really lives in the ideal netherworld of Mathematics with a capital 'M'.

Determining the value of π has been going on since the time of the ancient Babylonians and Egyptians, from 2,000 years B.C. right up to the present day. Why this is so, we will see shortly. The Babylonians used $3\frac{1}{8} = 3.125$ for the value of π, and the ancient Egyptians thought that a circle of diameter 9 had the same area as a square of side 8. Then, according to the formula for the area of a circle that we all learned in school, $A = \pi\, r^2$, this would mean that $\pi\, (4.5)^2 = 64$, giving a value of π of 3.1604. These are both reasonable values, when you consider that the State Legislature of Indiana in 1897 had a bill considering three different values for π, namely 3.2, 3.23, and 4. Fortunately for the reputation of the citizens of Indiana, the bill subsequently died a quiet death.

Since these ancient times, the value of π has steadily become more and more accurately known through the centuries. Anyone who wishes to pursue a fascinating account of this history can find it in Petr Beckmann's delightful book, *A History of Pi* (see Bibliography).

What is the point of pursuing more and more decimal places of π? There are trivial reasons of course, like seeing who can memorize the greatest number of digits. Years ago, the author decided — for no reason whatsoever — to memorize the first 20 decimal places of π which he can now show you:

$$\pi = 3.14159265358979323846\ldots$$

(It did impress some students however). He genuinely did not look these up to insert it here. But this is nothing compared to the current (as of this writing) unofficial record of 100,000 memorized decimal places of π, as rattled off by Akira Haraguchi of Japan, on October 3–4, 2006[37].

A more practical reason is that it is used as a test for supercomputers. If you already know that the quadrillionth digit of π happens to be a 6 (no one knows if it is or not, yet it exists already in the mathematical netherworld), then any new supercomputer can be put through its paces to see whether or not it finds a 6 at the quadrillionth decimal place.

---

[37] This is the unofficial world record, as the accepted official world record is 70,000 decimal places, set by Rajveer Meena of India on March 21, 2015.

There is one final reason: because it is there. On a serious note, some very useful mathematical work has arisen from the attempts to compute the humble π to phenomenal numbers of decimal places.

As it happens, the first 31,415,926,535,897 digits of π have recently been determined. The monumental task was completed on January 21, 2019, by Emma Haruka Iwao, using a program called 'y-cruncher' developed by Alexander Yee[38]. In fact, you can even download the y-cruncher program for yourself[39]. This program now has the facility to compute other constants to absurd numbers of decimal places, such as the exponential e and the golden ratio φ (both discussed later in the text), and several world records have been set with it in the process.

Just how *do* you compute the value of π to a trillion decimal places? The actual method used by Iwao is somewhat beyond the scope of this book, but it is essentially based on a remarkable discovery from the mysterious realm of infinite series that will be discussed at length in the next chapter[40]. What lies beyond the 31.415… trillionth digit of π? More digits of course, but at this moment in time no one knows what they are, although they are out there, in math-world, waiting to be discovered[41].

OFF TO MONTE CARLO

The very definite nature of π does not suggest that it could be accessible by any sort of random methods. But here our assumption would be wrong.

(i) Draw a square with each side having length 1 (unit). The length is really of no significance and any length would do. Clearly the area of this square is 1.

(ii) Inscribe a circle inside the square; that is, with the circle touching each side of the square as in Fig. 1.11. This means that the radius of the circle is $r = 1/2$. Then the area of the square, call it, A_s, equals 1 and the area of the circle is: $A_c = \pi (1/2)^2 = \pi/4$. Let us now consider the *ratio* of the area of the circle A_c ($= \pi/4$) to that of the area of the square A_s ($= 1$) and write down our results:

$$\frac{A_c}{A_s} = \frac{\pi/4}{1}.$$

---

[38] The project was started on September 22, 2018 and took 111.8 days of computer time.

[39] Crunch away at http://www.numberworld.org/y-cruncher/#Download

[40] The actual infinite series that was used is the Chudnovsky algorithm that can be found at: https://en.wikipedia.org/wiki/Chudnovsky_algorithm

[41] For those interested, the last 97 digits of Iwao's calculation of π are: 6394399712531109327 6981435565618400374993573460992143395529689721224771577728930842732326247390.

The right-hand side simply equals $\pi/4$ and this is now the ratio of the area of the circle to the area of the square. It will not be necessary to work out the value $\pi/4$ on a calculator; we want to leave it just as it is. But very roughly, the area of the circle is a bit over three-fourths the area of the square.

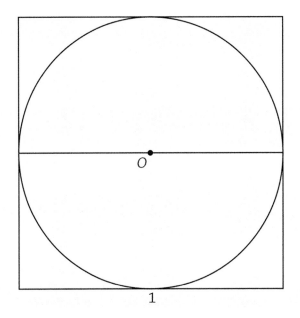

Figure 1.11: A square of unit length with an inscribed circle.

(iii) Now scatter at random a large number of some small particles, like grains of sand or rice, over the entire square. Some will land inside the circle and some will be in the square but not the circle. Ignore any grains that are right on the circumference of the circle or outside the square.

(iv) Now let us count the number of grains that lie inside the circle and call this number N_C, and say that N_S is the total number of grains lying within the square itself.

We now have all the ingredients to determine an approximate value of π.

Again, let us take a simple ratio, this time the ratio of the number of grains that landed in the circle to the total number of grains in the square; that is, N_C / N_S. Now, our mathematical intuition tells us (assuming we distributed the grains randomly) that this ratio of the number of grains that are in the circle to the number in the square should be the same as the ratio of the *areas* of the two figures. From the preceding calculation, we found this ratio to be equal to $\pi/4$. So, simply by counting grains we have determined that the grain ratio N_C / N_S, should in a very real sense be equal to the area ratio $\pi/4$. Writing this down we have,

$$\frac{N_C}{N_S} = \frac{\pi}{4}.$$

At this stage, we simply multiply the number N_C / N_S by 4, which gives us the expression for π,

$$4\frac{N_C}{N_S} = \pi.$$

This is just a simple numerical calculation, which we obtained by a very basic and random mechanical means. Of course, this is somewhat tedious. Using more grains of rice/sand should make the approximation more accurate, and doing it on a computer is easier still, but the point is, it works.

The idea at the heart of the random procedure described above is a very important simulation technique known as the *Monte Carlo Method* and was employed at Los Alamos in the Manhattan atom bomb project during World War II[42].

With the advent of computers, it has become a very important modelling tool in Physics, Chemistry, and Engineering and Applied Statistics. In our example, instead of physically taking grains of sand, we could have a circle inscribed in a large square on a computer screen and a simple program could choose pixels at random lying inside the square. Every time a pixel was chosen the program would simply keep a tally of whether the pixel was inside the circle. Let the program do this a few thousand times and then compute the ratio of hits inside the circle to that of total hits in the square, multiply by four and there we have it. We have made a determination of π by completely random means.

The author actually did the experiment via a computer simulation of pixels randomly placed on a square with an inscribed circle and got a value for π of at least 3.14 on most simulations, and sometimes a few more decimal places on others. The reader is invited to do the same.

SMASHED PI

A most intriguing and unusual method of determining the digits of π was discovered in 2003 by Gregory Galperin[43]. In this ideal scenario there are two billiard balls, having masses m and M respectively, on a frictionless flat table. The ball of

[42] The idea originates with Polish-American mathematician Stanisław Ulam (1909–1984), of whom we will hear more about in Chapter 5.

[43] G. Galperin, Playing pool with π (The number π from a billiard point of view), *Regular and Chaotic Dynamics*, 2003, pp. 375–394.

smaller mass m is positioned to the left of the larger mass M, with a fixed vertical wall to the left of the smaller ball as in Fig. 1.12.

Now we take a shot at the larger ball, setting it in motion directly at the smaller one. We will also assume that all collisions are perfectly *elastic*, in that no energy is lost to heat or sound etc. Atoms collide in this fashion. What happens next all depends on the value of the ratio of the two masses m and M, namely M/m.

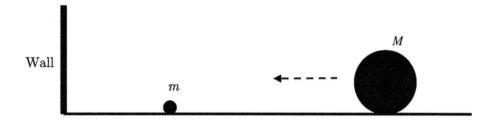

Figure 1.12: The setup of two billiard balls having masses m and M respectively, sitting on a frictionless table with a wall to the left. The wall acts as a barrier to reflect the balls back and forth once set in motion by the larger ball striking the smaller one, which is set away from the wall. The surface continues indefinitely to the right.

In this ideal elementary scenario, we are simply interested in the total number of collisions between the two balls themselves, plus the small ball with the wall. Nothing else.

Since the ratio of the two masses is the crucial factor here, let us take the logarithm of the ratio and give it a name:

$$N = \log_{100} \frac{M}{m}.$$

The base of 100 is only a technical artifice so we can just go along with it. Thus, we have transferred any dependence on the ratio M/m to the single number N[44].

To see what unfolds, let us start with a few simple cases and count collisions.

If the two masses happen to be equal, that is $M = m$, then their ratio equals 1 and $N = \log_{100}(1) = 0$. How many collisions are there? By the laws of Physics, when the right-hand ball in motion strikes the stationary ball (= 1 collision) it imparts all its velocity to the stationary one, hence it stops. The formerly stationary ball consequently will move to the left and strike the wall (= 2 collisions), bouncing

---

[44]This is the same as writing:

$$\frac{M}{m} = 100^N,$$

so that increasing the integer N by 1 increases the ratio of M/m by a factor of 100.

back to strike the other ball (= 3 collisions) and setting the latter on its way indefinitely to the right. So, when $N = 0$ we have a total of three collisions.

Instead of equal masses, let us up the ante a bit and make $M = 100 \times m$ so that the bigger mass is 100 times that of the smaller one. Then $M/m = 100$ and the number N becomes:

$$N = \log_{100} 100 = 1.$$

In this instance, it turns out that the total number of collisions will be 31. This can be verified via a computer simulation but it is also a consequence of what Galperin proved mathematically (see below).

Increasing the ratio of the masses still further, let us now take the mass of the larger billiard ball to be $M = 10,000 \times m$. This is one monstrous billiard ball, but this is only an idealized situation, so let us play on. In this case,

$$N = \log_{100} 10,000 = 2,$$

and it turns out that the total number of collisions is 314.

If we continue to increase the ratio of M to m systematically in powers of 100, then Table 1.1 reveals the results:

Table 1.1. By increasing the number of collisions by a factor of 100 each time (given by increasing N by 1), the total number of collisions yields the $N + 1$ digits of π.

N	Total number of collisions
0	3
1	31
2	314
3	3141
4	31415
5	314159
...	...

Now, these numbers should trigger off a light bulb moment. What Galperin *proved mathematically* was that:

For each integer value of N (that is, increasing the mass ratio M/m by a factor of 100 each time), the resulting total number of collisions are the first $N + 1$ digits of π.

This is all rather amazing, that colliding balls can somehow produce the digits of π. If you are thinking that it might have something to do the billiard balls being round, then you would be incorrect, as the argument works just the same if the billiard balls are replaced by sliding cubic blocks.

When Galperin first spoke about his result at a mathematical colloquium, no one at first believed it could be true. However, after explaining the proof, the audience became believers. This complete disbelief and then acceptance upon seeing the proof was repeated at many learned institutions where the result was presented[45]. So if the reader is in a state of disbelief, rest assured that many before you have had the same experience.

THE DIVINE ISOPERIMETRIC INEQUALITY

The circle is also a divine figure in another interesting respect, at least in our Universe. Suppose you have a length of string and have glued or taped the two ends together in order to form a closed loop. The rest is such a simple procedure that you can do the entire exercise mentally.

Now, take this loop of string and pull and prod it to make any sort of plane figure out of it that you wish, and note that the area inside the loop can take on many different forms. Infinitely many, if we may use that term. Now, you should look to arrange the loop of string with the sole condition that the area inside the loop should be as big as possible. Anyone can do this experiment, even a child. After a bit of experimentation, we can expect that you will come to the conclusion that the shape of the string which encloses the largest area is a circle (Fig. 1.13).

This is not only intuitively true, and experimentally looks true, but can be mathematically proven to be true, and it goes by the name of the 'Isoperimetric Inequality'. The 'inequality' part of this expression derives from the fact that the area of the circle is bigger than all the other competing areas; in other words, all the other areas will have a strictly smaller area than a circle.

Since the Isoperimetric Inequality involves the notion of length and area, let us denote the length of the string by L, and the area of the enclosed figure as A. If we make a circle with our piece of string, then the length L becomes the perimeter (circumference) of the circle. So, let us ask ourselves what, if any, is the relation between a circle's circumference L, and its area A? Most readers will probably not have given this issue any thought, even though most of us learned about the area and circumference of circles in school.

For one thing, we know that the circumference L of a circle of radius r is given by

$$L = 2\pi r,$$

---

[45] A wonderful implementation of the collisions using two blocks can be found at: https://www.youtube.com/playlist?list=PLZHQObOWTQDMalCO_AXOC5GWsuY8bOC_Y

and the area A is given by,

$$A = \pi r^2.$$

When learning these two facts at school, you probably never realized that they are actually related to one another. For one thing, they both involve the value of π, and for another they both involve the radius of the circle r.

Note, however, that the area A involves r^2 and the circumference L involves, simply, r. So let us take the square of L to make the two equations look more alike and see what happens:

$$L^2 = 4\pi^2 r^2.$$

If we compare the right-hand side of this expression with the equation for the area, $A = \pi r^2$, we see that they are nearly the same, except that we need to multiply A by 4π, to make them *exactly* the same; that is,

$$L^2 = 4\pi^2 r^2 = 4\pi A.$$

Thus, the area of a circle and its circumference are indeed related to one another by the above expression. This is the crux of the Isoperimetric Inequality, which states that all the other competing areas A, except for the circle, will have a *smaller* area given the same fixed perimeter length L. This means that that: $4\pi A < L^2$ for the other (non-circular) competing areas, since for the circle we have $4\pi A = L^2$.

In general, we can put both statements regarding the area of any enclosed plane figure into one expression. This results in the Isoperimetric Inequality statement that: Given a fixed perimeter length L, then the enclosed area A of a plane figure satisfies:

$$L^2 \geq 4\pi A, \tag{1}$$

with equality holding only for a circle.

It seems only natural that, since a circle comes into play, π should make an appearance here. What the equation is saying is that given a string of length L that makes any enclosed area A in the plane whatsoever, then the length and area of all such figures are related by the preceding inequality. Moreover, all competing areas will be strictly less than the value L^2 on the left of eq. (1) except for the figure of a circle for which $L^2 = 4\pi A$ in eq. (1). Therefore, the circle has the largest value of A of all the competing enclosed plane figures (Fig. 1.13).

So here, we have arrived at a fundamental property of our Universe, illustrated with a simple piece of string (not to be confused with String Theory). Of course, there are rigorous mathematical proofs of the Isoperimetric Inequality from various perspectives, because of the inequality's fundamental nature. Furthermore, if

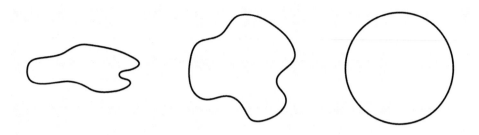

Figure 1.13: Three closed figures having the same length perimeter. The circle encloses the largest area, which is the conclusion of the Isoperimetric Inequality. (Illustration courtesy of Katy Metcalf.)

you were given a closed surface of material, like that of a deflated soccer ball, and were asked what shape of the material would enclose the largest volume, you would most likely guess that the answer would be a spherical ball and you would be exactly right[46].

The solution to the isoperimetric problem in its two- and three-dimensional forms was already known to the ancient Greeks.

There is a whole range of so-called 'isoperimetric inequalities' that "relates two or more geometric and/or physical quantities associated with the same domain. The inequality must be optimal in the sense that the equality sign holds for some domain or in the limit as the domain degenerates[47]."

There are many applications in Physics that fall within this genre. For example, one can lower the fundamental frequency (tone) of a drum merely by making the drum bigger, but what about drums of a given membrane size? It is known that for all (homogeneous) vibrating membranes with a fixed given area, it is the circular membrane that gives the lowest fundamental frequency. This was initially conjectured by Lord Rayleigh in 1877 and subsequently proved by others in the 1920s. So, something that is a basic cornerstone of the Mathematical Universe also has an elementary consequence in the world of music.

---

[46] In its three-dimensional version the Isoperimetric Inequality says that if S is the surface area of a closed three-dimensional region, and V its volume, then $S^3 \geq 36\,\pi\,V^2$, with equality holding only for a sphere. In other words, the sphere encloses the largest volume for any given surface area. The relevant formulas are: $V = \left(\frac{4}{3}\right)\pi\,r^3$, and $S = 4\pi r^2$. With a bit of algebra, one can verify that you have equality: $S^3 = 36\pi V^2$ in the case of a sphere.

[47] From: Isoperimetric inequalities and their applications, *SIAM Review*, 9, 1967, 453–488.

Here, we get a sense of what mathematicians refer to as 'beauty' in Mathematics. If the answers to the above two- and three-dimensional isoperimetric problems had been some bizarre figures other than a circle and a sphere, we would find this situation much less aesthetically pleasing or satisfying. But in general, in the mathematical world, often the most beautiful solutions are ones that are proven to be correct. In the words of famed 20th century German mathematician, Hermann Weyl, "My work always tried to unite the true with the beautiful, but when I had to choose one or the other, I usually chose the beautiful."

In the sciences, one must of course be careful of the lure of beauty. As Nobel Prize winning physicist Richard Feynman warned, "It doesn't matter how beautiful your theory is, it doesn't matter how smart you are. If it doesn't agree with experiment, it's wrong."

But of course, this admonition from Feynman must come with a caveat of its own. In 1957, physicist Murray Gell-Mann and some colleagues, "put forward a partially complete theory of the weak force in disagreement with the results of seven experiments. It [the theory] was beautiful and so we dared to publish it, believing that all of those experiments must be wrong. In fact, they [the experiments] *were* all wrong[48]."

Such is the winding path upon which Science advances.

---

[48] From a TED talk by Gell-Mann: *Beauty, truth and … physics?* 2007: https://www.ted.com/talks/murray_gell_mann_on_beauty_and_truth_in_physics#t-96491

2

From Here to Infinity

The Infinite! No other question has ever moved so profoundly the spirit of man...
Mathematician David Hilbert

The interior of our skulls contains a portal to infinity...
Writer Grant Morrison

When the author was a Mathematics student at the University of California, Los Angeles (UCLA), there was a mosaic on the sides of the Mathematics building containing various symbols. One of these was the number 3×10^{74}, which is a shorthand way of writing the enormous number 3 followed by 74 zeros. The author was told that this was the number of atoms in the observable Universe, although it has now increased over the years to roughly 10^{80}. The qualification 'observable' is important here and represents a sphere of about 93 billion light-years in diameter, though the actual shape of the entire Universe is still not known[1].

These big numbers are, of course, a very rough estimate, since such a calculation is necessarily difficult to make. But nevertheless, the number is finite although the Universe itself is possibly infinite. So, if the number of atoms in the observable Universe is finite, what then, if anything, in the known world is infinite? The answer is: possibly nothing. Certainly, in Physics, there are no measurable quantities that are infinite, although the term does arise. According to theoretical physicists Stephen Hawking and Leonard Mlodinow in their very readable book, *The Grand Design* (see Bibliography), Einstein's Theory of Relativity "predicts

---

[1] It should be noted that the number 3×10^{74} does appear in the study of phylogenetic trees and is the number of unrooted trees generated by 50 taxa, for those who know some Biology.

© Springer Nature Switzerland AG 2020
J. L. Schiff, *The Mathematical Universe*, Springer Praxis Books,
https://doi.org/10.1007/978-3-030-50649-0_2

there to be a point in time at which the temperature, density and curvature of the universe are all infinite, a situation mathematicians call a singularity. To a physicist this means that Einstein's theory breaks down at that point and therefore cannot be used to predict how the universe began, only how it evolved afterwards." Similarly, singularities have been associated with the center of a black hole where spacetime curvature is supposed to become infinite.

Mathematicians are very familiar with the term singularity, as this is a point where certain functions become arbitrarily large the closer we get to them. A simple example is the function,

$$y = \frac{1}{x^2}.$$

As the value of x gets closer and closer to zero, the value of y becomes arbitrarily large[2]. The value of the variable x can get as close as we wish to zero, but we must stay away from the point $x = 0$ itself, because when x is at that point, y simply becomes *undefined*. Here, the point $x = 0$ is a singularity for this particular function (Fig. 2.1).

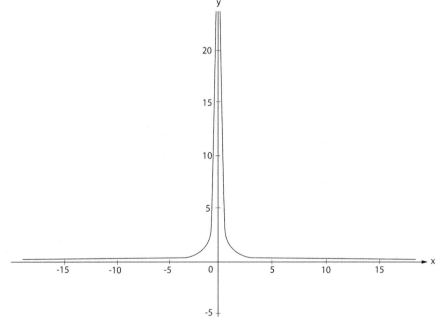

Figure 2.1: The graph of $y = \frac{1}{x^2}$ showing how the value of y becomes infinitely large as x approaches the value zero.

---

[2] We can say that y 'becomes infinitely large' if we understand that term to mean 'becomes arbitrarily large' as x gets closer and closer to zero. Also note that there is no real number that can be ascribed to y when $x = 0$. Nor is there a number called 'infinity', a point that will be reiterated later in the text.

As Brian Greene observes in his bestselling, *The Elegant Universe*, "Like a sharp rap on the wrist from an old-time schoolteacher, an infinite answer is nature's way of telling us that we are doing something that is quite wrong."

The very notion of something infinite has fascinated the imagination of mankind through the mists of time. There are of course the mathematical notions of infinity, which we will discuss later in the text, but there are many others as well. Many of us use it in every day discourse simply to indicate something enormous, as in this musing by French philosopher Voltaire: "The only way to comprehend what mathematicians mean by Infinity is to contemplate the extent of human stupidity." Albert Einstein said something similar[3].

But there are other connotations about the infinite as being something deeply mysterious about life itself beyond common experience, as in the Zen proverb, "The infinite is in the finite of every instant." One of the very aims of this book is to explore and examine some of the mystery of the world we inhabit as embraced in this proverb. To others, the infinite has religious overtones of an eternal God of unlimited, omnipotent qualities, who is separate from our mundane, finite world yet also a part of it. To most people, the Universe itself is something that is infinite, something impossible to conceive of ever ending. But even here the jury is out, as will be discussed in Chapter 6.

ZENO'S PARADOX

What follows is a classical example, known as the 'Dichotomy', that appears in discussions of the infinite. It was proposed by the 4th century B.C. Greek philosopher Zeno of Elea. Unfortunately, Zeno's various paradoxes are only known second hand through the works of Aristotle (384–322 B.C.) and the pagan philosopher Simplicius (*ca.* 490 – *ca.* 560). Even if you have seen this paradox before, you have quite possibly forgotten how to resolve it. From this simple example, you will immediately be able to grasp the mysterious quality of the notion of infinity. Zeno proposed other paradoxes having to do with motion, in an attempt to discredit the very concept of motion and thus the illusory world of our senses.

Here is what Zeno had in mind. Suppose you wish to take a walk to the corner store. At some instant in time, you will be exactly at the half-way point (at least some point on your body will be). Then, as you walk further, at some instant you will pass the exact half-way mark of the remaining distance, and again at some instant you will pass the half-way mark of the distance that still remains, and so on. This is going to require your body to cover an infinite number of finite

[3] "Two things are infinite, the Universe and human stupidity, and I am not yet completely sure about the Universe." In various forms, this has been attributed to Einstein, but the quote is only known second hand through books by psychiatrist Frederick S. Perls.

distances and so you cannot complete the journey. Indeed, even completing the first half of the journey requires going through the same procedure and so your trip to the store can never even take place. A pertinent related question is whether time itself is infinitely divisible; that is, like the continuum describing all the real numbers on a line.

Another variation is that of the 'arrow paradox', whereby an arrow that is shot into the air is, at any given instant of time, at rest at a specific position in space. Thus, there can be no motion of the arrow as it is at rest at every instant.

One interesting aspect affecting Zeno's paradoxes is the notion of an 'instant'. This is really a mathematical construct, in the same way that a 'point' or a 'continuum' is a mathematical construct. What can this mean for the nature of the physical world?

Therein lies the mystery: how close to physical reality does the mathematical instant come? According to Peter Lynds, "[T]here is not a precise static instant in time underlying a dynamical physical process at which the relative position of a body in relative motion or a specific physical magnitude would theoretically be precisely determined." If such a hypothetical instant did exist, then "... events and all physical magnitudes would remain frozen static, at such a precise static instant in time[4]."

In other words, there can never actually be a mathematical instant of time and *real* time is always an interval, much like a 'point' in space would have a three-dimensional volume.

Just to add a modern-day complication to this scenario, physicists do not even know if it is possible to divide space into ever smaller and smaller fragments, and there might even be a limit beyond which space becomes granular and is no longer divisible. According to theoretical physicist Carlo Rovelli, "There is a lower limit to the divisibility of space. It is at a very small scale indeed, but it is there."

This lower limit is considered to be on the order of the *Planck length* λ_p of 10^{-35} meters.

At this stage it should be mentioned that Max Planck (1858−1947), whose name appears scattered throughout this book, was a distinguished German physicist who won the 1918 Nobel Prize in Physics for work related to his discovery that electromagnetic energy is emitted in discrete packets that he called *quanta*. This was stated in his eponymous equation (eq. 25) found in Chapter 7. This fundamental discovery helped usher in the important field of Quantum Mechanics that will be discussed in that chapter.

It is even quite possible that time itself suffers from the same fate and that, at some minute scale, it too becomes discrete and is no longer divisible.

Certainly, in the period up to 10^{-44} seconds, (known as *Planck time*, t_p) after the Big Bang (Fig. 2.2), nothing can be known as it lies beyond the scope of our current Physics.

---

[4] Peter Lynds: Time and Classical and Quantum Mechanics: Indeterminacy Versus Discontinuity, *Foundations of Phys. Lett.*, 16(4), pp. 343−355, 2003. Note the recent date. Zeno's ideas are still with us.

Indeed, Planck time is the time it takes for a photon of light to travel the distance λ_p. Since distance equals velocity multiplied by the time,

$$\lambda_p = c\,t_p,$$

where c is the (constant) velocity of light[5].

Never before has a simple stroll to the store or shooting an arrow into the air been as fraught with such complexity.

Figure 2.2: It all started with the Big Bang, but what took place in the initial 10^{-43} seconds, is currently unknowable. (Image from a painting owned by the author.)

---

[5] Putting in some values, $\lambda_p = 1.62 \times 10^{-35}\,m$, and taking the velocity of light as $c = 3 \times 10^8\,m/sec$, gives a value for the Planck time,

$$t_p = \frac{\lambda_p}{c} = \frac{1.62 \times 10^{-35}\,m}{3 \times 10^8\,m/sec} = 5.4 \times 10^{-44}\,sec \sim 10^{-43}\,sec.$$

We will make use of the notion of discrete time (not necessarily Planck time), in our discussion of Cellular Automata in Chapter 5. However, in this latter case, the time will be represented by the tick of a clock, not the interval between ticks.

Some mathematicians tend to resolve Zeno's paradox with a glib argument involving infinite series, which will be discussed in detail shortly.

That Zeno's paradoxes are still topics of scientific discussion nearly 2,500 years later is because the exact nature of time and space is still not completely understood. The paradoxes are a testament to Zeno's brilliance.

The history of the notion of the infinite is a long one and the interested reader is directed to the excellent book *To Infinity and Beyond – A Cultural History of the Infinite* by Eli Maor (see Bibliography). The symbol for infinity is the lemniscate: ∞, which nearly everyone will recognize but almost no one understands. To clear the air right from the start, the author stresses that in a mathematical sense, ∞ is *not* a number, it is merely a symbol and nothing more. It is used as a mathematical concept, as we will see shortly.

But to a mathematician, the notion of infinity is very real, and it arises in many interesting scenarios. The most obvious is when discussing positive integers like 1, 2, 3, 4, 5, [6]... What exactly do the traditional three dots (ellipsis) mean after the last number? It means we keep on going *ad infinitum*, as it were. If we were to consider the collection or set of all of these positive integers, namely,

$$\mathcal{P} = \{1,\ 2,\ 3,\ 4,\ 5,...\}$$

we would say that the set was infinite, or that the set had an infinite number of members. We will make this statement more precise in a moment.

As life is finite, we cannot ever count all the members of this infinite set because we all know that the numbers in this set simply get larger and larger and larger without end. So, the set of positive integers, in spite of our inability to name all its members, exists somewhere in the netherworld of mathematical reality.

If we select out of this set just the *even* integers, that is, those divisible by 2, such as 2, 4, 6, 8, 10, … then we see that the even numbers also carry on indefinitely without end, and so the set of even numbers,

$$\mathcal{E} = \{2,\ 4,\ 6,\ 8,\ 10,...\}$$

is infinite as well.

But now comes the part that makes the notion of infinity so fascinating from a mathematical perspective. The second set of even integers is just a *part* of the first set of positive integers. We have left all of the *odd* ones out of the set of positive integers, namely,

$$\mathcal{O} = \{1,\ 3,\ 5,\ 7,\ 9,...\},$$

and, as you can see, this is an infinite set as well.

---

[6]The author is avoiding using the term 'natural' numbers, as some sources include the number 0 and some do not. Likewise, the term 'whole' numbers can include all the integers or just the numbers 0, 1, 2, 3, …. depending on the reference. So, let us stick with the term 'positive integers'.

Nevertheless, and this is the mysterious part, the set of even (positive) integers $\mathcal{E}$, and the set of all the positive integers $\mathcal{P}$, have the *same number of elements* even though the former is only a part of the latter. The same holds true for the set of odd (positive) integers $\mathcal{O}$, which also has the same number of elements as the set of positive integers $\mathcal{P}$.[7]

To appreciate this bizarre point, we have to make it explicitly clear what we mean by 'have the same number of elements', since we already know that we cannot possibly count all the members of an infinite set.

Here is how we will approach this issue mathematically. Suppose we had a classroom full of boys and girls and we made the claim that there were an equal number of both sexes. One way to prove this is simply to count the numbers of each sex and see if they are the same. Suppose we made the same claim that there was an equal number of boys and girls for the entire school, consisting of several thousand students. Physically counting them could be somewhat laborious, but there is another way. We could assemble all the boys and girls outside on the school grounds and ask each boy and girl to pair off with one another and stand as a couple together. Each boy would be paired with one girl and likewise each girl would be paired with one boy. If there was a single boy or girl left with no partner, then we would know that the number of boys and girls were *not* the same. But if everyone had a partner, then we would know immediately that the numbers of boys and girls *were* the same.

It is this notion of pairing that we apply to infinite sets, since they cannot be physically counted but they can be physically *paired*. All we have to do with the set of positive integers $\mathcal{P}$, and the set of even integers $\mathcal{E}$, is simply to construct a pairing between their members. For instance, we can pair each positive integer with its double:

$$1 \leftrightarrow 2;\ 2 \leftrightarrow 4;\ 3 \leftrightarrow 6;\ 4 \leftrightarrow 8;\ 5 \leftrightarrow 10,$$

and so on... In general, $n \leftrightarrow 2n$, for each positive integer n.

From this pairing (called a *one-to-one correspondence*), we conclude that the set of positive integers $\mathcal{P}$, and the set of even integers $\mathcal{E}$, *have the same number of elements*. In fact, we could equally have paired off the positive integers $\mathcal{P}$, with the set of odd integers $\mathcal{O}$, and reached the same conclusion[8].

Thus, remarkable as it seems, the sets $\mathcal{P}, \mathcal{E}$, and $\mathcal{O}$, all have the same number of elements, although the even and the odd integers, when taken collectively, comprise the set of positive integers. This contradicts our very basic notion regarding

---

[7] In general, even and odd integers also include their negative counterparts, but we will only be dealing with positive integers when using the terms 'even' and 'odd'.

[8] We can pair the positive integers $\mathcal{P}$ with the odd integers $\mathcal{O}$ via the correspondence:

$$1 \leftrightarrow 1;\ 2 \leftrightarrow 3;\ 3 \leftrightarrow 5;\ 4 \leftrightarrow 7;\ 5 \leftrightarrow 9,...$$

In general, $n \leftrightarrow 2n - 1$, for each positive integer n.

finite sets, in which the whole is greater than any of its constituent parts. For infinite sets, this rule does not apply, and a set can have the same number of elements as a proper subset of itself, as we have just seen.

The type of infinite sets exhibited by the positive integers, even integers, or odd integers, or in fact any set that can be paired as we have done above with the positive integers, is actually a certain kind of infinity.

When we refer to any number, say the number 7, then it can have two interpretations. It can represent the whole number following 6, in which case it is indicative of order and so is referred to as an 'ordinal'. But if it is intended to indicate quantity, for example 7 girls, then it is interpreted as a 'cardinal'. The word *cardinality* means number quantity, and this is now significant as we are interested in sets having various numbers of elements.

We will say that two sets (finite or infinite, it does not matter) 'have the same number of elements' if the members of the respective sets can be paired so that each member of both sets has a unique partner (i.e. no sharing of partners), and no member of either set is left without a partner. So, for example, the set consisting of the letters of the English alphabet, $\{a, b, c, d, …, z\}$, can be paired with the set of numbers $\{1, 2, 3, 4, …26\}$, or in fact, with any set of 26 objects.

This is exactly what we would mean in ordinary English if we said that two bridge clubs 'had the same number of members'; that is, each member of one club could be paired with exactly one member of the other club, and vice versa. We can also use the same notion to say that both groups (or sets) have the same cardinality. As we have seen above, the advantage of this pairing notion is that it allows us to compare the cardinality of a set of objects with the cardinality of another set without explicitly going through the counting process.

The upshot of all this discussion is that what we have derived above, by the pairing of the set of positive integers $\mathcal{P}$ with the set of even integers $\mathcal{E}$, is that both sets have the same number of elements. In math-speak, we say that both sets have the same cardinality and such sets are deemed *countable*. The same is true of the set of odd integers $\mathcal{O}$. Indeed, all three sets, $\mathcal{P}$, $\mathcal{E}$, and $\mathcal{O}$, have the same cardinality; that is, the same number of elements.

But what should we designate as the cardinality of these three sets of numbers? We need some symbol to represent the infinite cardinality of the number of positive integers. You might think that the symbol ∞ would be the one to use, but there are reasons why this symbol is not at all suitable, as we will see.

The cardinality of the set of positive integers $\mathcal{P}$, and likewise the set of positive even integers $\mathcal{E}$ and the set of positive odd integers $\mathcal{O}$, is denoted by, $\aleph_0$, which is called '*aleph-naught*' or '*aleph-null*'. The 'aleph' part is the first letter of the Hebrew alphabet. You might think that something representing the last letter of an alphabet would be more appropriate for the cardinality of an infinite set like the positive integers, but actually $\aleph_0$ represents only the first of many infinite cardinals. We are only getting started, but before we get to the others, let us try out some arithmetic with our new cardinal. We have all done arithmetic at primary school, but you have not seen arithmetic like this before.

Suppose we combine the set $\mathcal{E}$ (which has cardinality $\aleph_0$) with the set $\mathcal{O}$ (which also has cardinality $\aleph_0$) into one set. We end up with the set of positive integers $\mathcal{P}$, and this of course has cardinality $\aleph_0$ too. Expressing this mathematically, we have illustrated the remarkable fact that

$$\aleph_0 + \aleph_0 = \aleph_0. \tag{2}$$

That is, we have combined two infinite sets together (the set of positive odd and even integers), but the resulting set (the positive integers) has exactly the same number of elements as each of the two constituent parts that compose it. What is this? A set that has the same number of members as a part of itself? This certainly does not happen with finite cardinals: $7 + 7 = 14$. Okay, if you wish to be pedantic, $0 + 0 = 0$, but that is the only one.

This also means that we can throw in all the negative integers (and 0) into our set $\mathcal{P}$ and the resulting set of *all* the integers will still have cardinality $\aleph_0$.

We have just entered the intriguing realm of the theory of infinite sets, in the main developed by Georg Cantor, and the above expression is an example of his 'transfinite arithmetic'[9]. Now you can appreciate why so many of Cantor's contemporaries were absolutely baffled by this mysterious world of infinite sets, with some, like his teacher Kronecker, being outright hostile.

Cantor is also the one who introduced the aleph notation, as he may have been of Jewish heritage although this is not certain. If at first you feel a bit uneasy about expressions such as eq. (2), it must be pointed out that many of Cantor's contemporaries were none too thrilled with them either. In fact, many leading mathematicians of the time were highly scornful of Cantor's work on transfinite numbers and he was called a "scientific charlatan" and his work scorned as "utter nonsense", among other derogatory remarks. However, in time the mathematical world came to realize that what Cantor had created was something of great genius and many honors and awards were subsequently bestowed upon him. As the renowned German mathematician David Hilbert was to say some years after Cantor's death, "No one shall expel us from the Paradise that Cantor has created."

As it turns out, it does not matter if we add $\aleph_0$ to itself just once, as in eq. (2), or infinitely many (that is, $\aleph_0$ times), we nevertheless obtain

$$\aleph_0 + \aleph_0 + \aleph_0 + \cdots = \aleph_0.$$

---

[9] Infinite sets and some of their strange properties were previously considered by both Bernard Bolzano and Charles Sanders Peirce, but it was Cantor who most completely developed the theory.

Just as in ordinary arithmetic, if you add up 10 of the same number, this is the same as multiplying the number by 10. The same occurs for the preceding expression, so that we can re-write it as,

$$\aleph_0 \times \aleph_0 = \aleph_0.$$

Although it would seem that there is no getting 'beyond' $\aleph_0$, there is an even larger cardinal number, and in fact, an infinite hierarchy of infinite cardinals.

We have already discussed the whole numbers, but of course we know that there are many more. As we have seen, π is not a whole number, nor are fractions (except when the numerator is divisible by the denominator). There are other non-whole numbers, like $\sqrt{2} = 1.414\ldots$ or $\pi = 3.14159\ldots$, which simply cannot be expressed as a fraction, and hence are called *irrational* numbers. Fractions, being nicer in some sense, are known by the more erudite term *rational* numbers. If we take all the rational numbers, including all the positive and negative ones[10], and all the irrational numbers and make a set out of them, then we have a collection of numbers which can no longer be put into a one-to-one correspondence with the whole numbers. This is the set of *real* numbers $\mathbb{R}$ which form all the numbers on the familiar number line.

This set of real numbers is 'bigger' in the sense of cardinality than the set of all the integers (positive, negative, and zero), as was proved by Cantor with a very ingenious but simple argument (see Appendix III). The cardinality of this set is denoted by the letter $\mathfrak{c}$, which stands for the word *continuum* since all the real numbers constitute a continuous straight line. There are no gaps on this line, with every point on it being occupied by a real number, be it rational or irrational. In the language of transfinite arithmetic, we have the expression

$$\aleph_0 < \mathfrak{c},$$

which says that the cardinality of the all the integers is less than the cardinality of the real numbers. We deal with whole numbers and lines in our everyday life, and now we know that they represent *two different types of infinities and that one is bigger than the other.* How about that?

In actual fact, we do not even need the entire extent of the real numbers as represented by an infinite line in each direction. Any length of line will do − one that is a mere centimeter long will suffice. This is because any line, no matter how long or short, has exactly the same number of points, and that number is $\mathfrak{c}$.

---

[10] The rational numbers include all the integers, as any integer n can be written as a fraction: $n = n/1$.

Here is the reason why. We can take any two lines of different lengths, as in Fig. 2.3. Then you can see that for each point on the shorter line AB, we can pair it up with a unique point on the longer line $A'B'$ (e.g. C gets paired with C'). Conversely, any point on the longer line can be uniquely paired with one on the shorter one (e.g. D' gets paired with D). This means that both lines must have the same number of points, which is $\mathfrak{c}$.

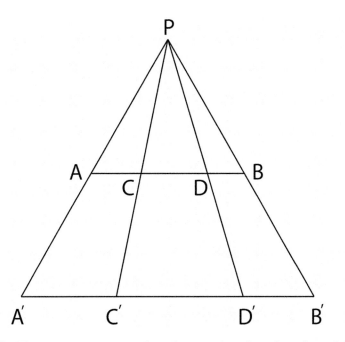

Figure 2.3: The one-to-one correspondence between the points of two lines of differing lengths ($C \leftrightarrow C', D \leftrightarrow D'$). This shows that both lines have the same number of points, $\mathfrak{c}$. (Illustration courtesy of Katy Metcalf.)

We can extract some more interesting arithmetic from the fact that any two lines have the same number of points. Take one line (with finite length) with $\mathfrak{c}$ points and attach one end of it to an end of some other line (of finite length) which also has $\mathfrak{c}$ points. This produces a longer line, but it too must have $\mathfrak{c}$ points; that is,

$$\mathfrak{c} + \mathfrak{c} = \mathfrak{c}.$$

Or, take the real numbers between 0 and 1 and combine them with the real numbers between 1 and 2, the real numbers between 2 and 3, the real numbers between 3 and 4, and so on. Essentially, attach intervals of the number line and adjoin them lengthwise. How many such intervals of real numbers have we adjoined? The answer is we have joined up $\aleph_0$ intervals each with c points, and produced yet another line, the right-half of the real number line, which still has c points, as all lines do.

Therefore, in the language of transfinite cardinals,

$$\aleph_0 \times c = c.$$

Here, we have actually multiplied two different infinities together and obtained a meaningful answer. We can also multiply $c \times c$, which would represent the number of points in a square, and the result would again be c. This is not at all obvious, and even Cantor originally did not think it possible to set up a one-to-one correspondence between the points of a square and the points on a line. In the end he succeeded, and the basics of the proof are shown in Appendix IV.

Is there any set having a cardinality between $\aleph_0$ and c? Cantor came to think there was no such cardinal but was unable to prove this. The problem became known as the 'Continuum Hypothesis' (CH) and many attempts to prove it ended in failure. In 1940, Austrian mathematician Kurt Gödel demonstrated that within the constraints of set theory, of which transfinite arithmetic is a part, the negation of the Continuum Hypothesis – that is, that there *is* a set with cardinality between $\aleph_0$ and c – cannot be proved. Not because it is too difficult, but because it cannot be proved *in principle*.

Then in 1963, American mathematician Paul Cohen proved that in set theory, it is not possible to prove the Continuum Hypothesis itself, again in principle. Therefore, the Continuum Hypothesis cannot be proved nor disproved even in principle, illustrating that there are certain limits to exactly what can be proved to be true in Mathematics, and what cannot. While this may seem only germane to the realm of very theoretical mathematical considerations, it is actually very relevant to the down-to-Earth subject of machine learning, as will be discussed at the end of Chapter 8.

There is another interesting proposition, known as the Axiom of Choice, that likewise cannot be proved nor disproved from the axioms of set theory[11]. That is why it is considered an axiom. It basically says that given a collection of sets (that are non-empty), one can find a procedure (known as a *choice function*) that picks out one element from every set. Sounds simple enough does it not? It is like going around your neighborhood when the trash collection is due and taking the topmost piece of trash out of each neighbor's bin. That is your choice function, assuming that is your inclination.

---

[11] The standard axiomatic formulation of set theory is due to Ernst **Z**ermelo and Abraham **F**raenkel and, coupled with the Axiom of **C**hoice (or an equivalent), is referred to as ZFC.

The issue becomes problematic when dealing with infinite sets. Suppose we consider all possible subsets of real numbers:

$$(0, 1), \{1, 2.6, 3.789, \pi\}, \{12\}, \{99, 17.43, 64.123\}, (100, 101), \ldots$$

where the first and fifth sets listed represent the intervals on the number line of *all* the real numbers *between* 0 and 1, and 100 to 101, respectively, not including their endpoints. Note that these two sets do not contain a largest or smallest number. So, we cannot just blithely pick the largest or smallest number out of each set, which would have been a tempting choice function and sometimes works with sets of integers. Nor can an arbitrary number be selected from each set, since this procedure gives no indication of what the number is to be and is thus not a suitably defined choice function. Now the reader can see that the axiom is not so obvious any longer.

It must be said that most mathematicians accept the Axiom of Choice as being true in their daily work. On the other hand, it does have a very interesting consequence which is slightly unpalatable, namely the *Banach-Tarski paradox*. This latter states that a solid ball in three-dimensional space can be cut up into a finite number of pieces, and then the pieces reassembled only by translating and rotating them (no stretching or distortion), in order to create *two* balls *identical* to the original. At this juncture the reader is allowed to utter, "Whatever!"

On the other hand, mathematical physicist Bruno Augenstein has found links between the 'paradoxical' decompositions in the Banach-Tarski theorem and the quantum theory of quarks, discussed in Chapter 7[12]. Could it be that this bizarre fact of the Mathematical Universe has a counterpart on some level in the quantum universe?

So, here we have had a taste of the arithmetic of the infinite. It clearly does not behave like the arithmetic we all learned in school, but perhaps we should not expect it to. We are dealing with infinite sets after all, not 6 and 7, so we should expect a few surprises.

If you have some neurons that are still firing properly, we can just mention that the cardinal number c turns out to be exactly the same as the cardinal number $2^{\aleph_0}$; that is,

$$2^{\aleph_0} = c,$$

where $2^{\aleph_0}$ represents the number of all possible subsets of whole numbers, like $\{0, 1, 2, 3\}$, $\{17\}$, $\{6, 11\}$, etc. This has a certain aesthetic appeal in the sense that the first transfinite cardinal is $\aleph_0$ and (assuming the validity of the Continuum Hypothesis) the second transfinite cardinal is $2^{\aleph_0}$.

---

[12] B.W. Augenstein, Hadron physics and transfinite set theory, *International J. Theoretical Physics*, 12, 1197−1205, 1984.

But a hierarchy of infinities was mentioned earlier, an infinity of infinities. Briefly, we can raise 2 to the power of $2^{\aleph_0}$, which yields $2^{2^{\aleph_0}}$, and represents the (cardinal) number of all possible subsets of real numbers. If we continue to raise the number 2 to the power of the previous cardinal, we will continue to get larger and larger cardinals... There we have it, an infinity of infinities.

SUMMING UP

We all know how to add numbers, but how do you add an infinite number of numbers? Mathematicians have devised a very intuitively natural procedure for doing this. Infinite sums of numbers, known as *infinite series*, come into daily life every time you pick up your cellphone or turn on the TV, as they are at the very heart of digital signal processing. In such instances, the original continuous wave signal is sampled at many discrete time instants per second and then transmitted in a digitized format. At the receiving end, your cellphone or TV then reconstructs the original signal via an infinite series, as if by magic[13].

Our understanding of the nature of waves derives from the introduction of *Fourier series* by French mathematician John-Baptiste Joseph Fourier in 1822. This is a special type of infinite series involving sine and cosine functions. Fourier series are also used to find solutions to, for example, the *heat equation* that mathematically describes heat flow in a solid object[14]. It would be fair to say that life as we know it would not be possible without a deep understanding of infinite series. Fortunately for our purposes, we do not need to go very deep.

Suppose we have an infinite collection of numbers, which we denote by:

$$a_1, a_2, a_3, a_4, \cdots$$

and we wish to find their sum. Then we would write the sum like this:

$$a_1 + a_2 + a_3 + a_4 + \cdots$$

calling this an infinite series.

But life is short and there are infinitely many numbers to add up, so we cannot physically add this sum even on the fastest computer, as the sum never ends. What can we do?

---

[13] Such as the *Whittaker-Shannon interpolation formula* https://en.wikipedia.org/wiki/Whittaker–Shannon_interpolation_formula/ among various others.

[14] Solving the heat equation was in fact Fourier's initial reason for introducing his series. *Fourier analysis* is found wherever waves are involved, such as in signal processing (mentioned above), Acoustics, Digital Image Processing, Electrical Engineering, Optics, and even Quantum Mechanics.

Our only hope is to find a number S representing the infinite sum by some means *other* than repeated addition *ad infinitum*. So, let us begin in a very natural way, namely adding one number at a time but also taking note of the sum at each stage.

Here is what we do. Starting with the first term, a_1 of our infinite series, let that be our first sum s_1 just for bookkeeping purposes.

The sum of the first two terms is: $a_1 + a_2 = s_2$;

The sum of the first three terms is: $a_1 + a_2 + a_3 = s_3$,

and so on, so that sum of the first n terms is:

$$s_n = a_1 + a_2 + a_3 + \cdots + a_n.$$

Taking n larger and larger is equivalent to adding more and more terms in our ever-expanding finite sum.

These finite sums are called the 'partial sums', as each one is summing up part of the original series of numbers. Each of these partial sums is a sum of finitely many numbers and so is easily doable. But of course, we cannot and do not do this indefinitely.

The mathematical ingenuity in all of this is to consider the values of the partial sums s_n as n gets progressively larger and larger; that is, by adding up ever more terms in our sum.

Suppose that by adding more and more and more terms to our finite sum (one at a time), the value of the sum at each stage s_n becomes closer and closer to some value S. If this happens to be the case, we designate S as the value of the infinite sum. Thus, we have achieved our aim of determining a value for an infinite series without carrying out an infinite number of additions, which is of course impossible.

This procedure should appeal to the reader's intuition as a natural thing to do: add up more and more terms and see if the partial sums s_n are approaching some value the more terms you take. If it does, then that is the natural value to take for the infinite sum. As will be seen subsequently, we will explore other ways to define the sum of an infinite series, but this is the most straight forward and common way.

If the sum of the infinite series $a_1 + a_2 + a_3 + a_4 + \cdots$, is the value S, in the above sense, then we write

$$S = a_1 + a_2 + a_3 + a_4 + \cdots$$

This approach is only valid whenever it is possible to find the limiting value S of the partial sums s_n, but for the most part, those are the most interesting infinite series that scientists have to deal with.

Such infinite series are said to be *convergent*, since the partial sums s_n *converge* to the sum S as n becomes arbitrarily large. We can write this in a mathematically succinct way as:

$$s_n \to S \text{ as } n \to \infty.$$

The first term '$s_n \to S$' means that the values of s_n approach the value S, and '$n \to \infty$' simply means that the number n becomes larger and larger *ad infinitum*. We have already mentioned that there is no such number as ∞ and so taking whole numbers n *ad infinitum* is a mental construct. Joining the two conditions as above means that the former occurs in conjunction with the latter.

Moreover, the sum value S is called the *limit* of the partial sums s_n, and this limit S represents the sum of the infinite series whenever it exists. Of course, we still have to find this limiting value S that the partial sums s_n converge to, but this is often possible, although not always. It is time to look at an example of how this all works in practice.

Let us consider the important *geometric series*,

$$\frac{1}{2} + \frac{1}{4} + \frac{1}{8} + \frac{1}{16} + \cdots + \frac{1}{2^n} + \cdots$$

and its partial sums s_n. Here are the first few:

$$s_1 = \frac{1}{2};$$

$$s_2 = \frac{1}{2} + \frac{1}{4} = \frac{3}{4};$$

$$s_3 = \frac{1}{2} + \frac{1}{4} + \frac{1}{8} = \frac{7}{8};$$

$$s_4 = \frac{1}{2} + \frac{1}{4} + \frac{1}{8} + \frac{1}{16} = \frac{15}{16};$$

$$\vdots$$

and so forth[15]. It is pretty clear, and this is established rigorously in Appendix V, that the partial sums s_n are approaching the value 1 as n becomes larger and larger.

---

[15] In general, the *nth* partial sum can be determined by employing a bit of algebra (see Appendix V for details):

$$s_n = 1 - \frac{1}{2^n},$$

for $n = 1, 2, 3, 4, \ldots$, and can be easily verified in each of the above cases for s_n.

Therefore, 1 is the value of the sum of the geometric series, i.e.,

$$\frac{1}{2}+\frac{1}{4}+\frac{1}{8}+\frac{1}{16}+\cdots+\frac{1}{2^n}+\cdots=1.$$

There we have it. We have just added up one of the most important infinite series in all of Mathematics by using a simple but clever device involving only the consideration of the partial sums of the series and subsequently determining their limit. Such are the wonders of Mathematics that we have achieved something that is impossible to do in real life, yet has ramifications for real life.

A visual demonstration of this is given in the image of a unit square divided up according to all the terms in the geometric series, in Fig. 2.4.

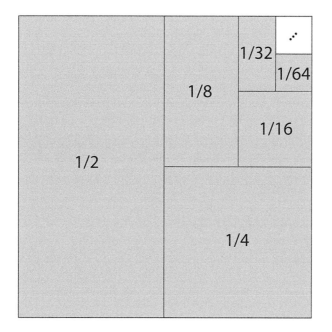

Figure 2.4: Graphical depiction of the unit square divided up according to the terms of the geometric series. The total area is tending to the value 1. (Illustration courtesy of Katy Metcalf.)

This particular series has a bearing on the Zeno Paradox mentioned previously. If it takes half an hour to complete the first half of your journey, and a quarter of an hour to complete the next half, and an eighth of an hour to complete the next half, and so on, then the journey should be complete in 1 hour. However, Zeno's Paradox has turned a question about the nature of the physical world into a mathematical one, and as we have mentioned, it may not be possible to divide time and space into ever decreasing quantities.

Here, the real world brushes hard up against the ideal mathematical world. "Our belief that the mathematical theory of infinity describes space and time is justified to the extent that the laws of physics assume that it does, and to the extent that those laws are themselves confirmed by experience[16]."

So yes, you can walk to the corner store and arrows do fly, although there are a multitude of explanations as to why this is so.

Similarly, it is known that

$$\frac{\pi}{4} = \frac{1}{1} - \frac{1}{3} + \frac{1}{5} - \frac{1}{7} + \frac{1}{9} - \cdots \tag{3}$$

which is a result attributed to the Indian mathematician-astronomer Mādhava in the 14th century and to James Gregory in 1671, and is also known as the *Leibniz series*. How in the world can π, which is the ratio of the circumference of any circle to its diameter, have anything whatsoever to do with adding up a series of fractions with all the denominators being odd numbers? Well it does, but it seems simply too good to be true.

There is one other point to make here. As we have seen, the number π is an infinite decimal expression that never ends nor infinitely repeats, and so it can only exist in this form in the mathematical realm. Likewise, the infinite series expression never ends, and so it, too, must exist only in the netherworld of Mathematics. But in that world, the two expression are precisely equal.

When the partial sums s_n do not get closer and closer (*converge*) to some finite number with increasing values of n, we say the series *diverges* or is *divergent*. We will see some instances of divergent series shortly.

You might well ask just how one adds up such an infinite series as the Mādhava-Leibniz series; that is, what value do the partial sums s_n converge to? Eq. (3) says that the partial sums converge to $\pi/4$ and indeed, the result was known long before the advent of computers. The proof involves the series representation of a certain elementary function and a bit of algebra[17]. The details need not concern us here, although the result is rather beautiful.

---

[16] *Stanford Encyclopedia of Philosophy*: https://plato.stanford.edu/entries/paradox-zeno

[17] Briefly, it can be shown that the function $y = \arctan x$ can be represented by an infinite series; that is, $\arctan x = x - \frac{x^3}{3} + \frac{x^5}{5} - \frac{x^7}{7} + \frac{x^9}{9} - \cdots$ Eq. (3) follows by noting that for $x = 1$, $\arctan(1) = \frac{\pi}{4}$.

But here is a related idea. Suppose we do add up the first million terms using a computer program, what would we expect to get? We would obtain an approximation to $\pi/4$, and the more terms we add up, the better the approximation would become. In fact, all we ever use in practice is an approximation to π anyway, depending on the requirements. Sometimes 3.14 will be all we need, and generally 3.14159 is quite sufficient for most purposes. So, an approximation that is correct to five or six decimal places is good enough. In fact, any calculation done on a computer is going to be constrained to a finite number of decimal places anyway.

The author took the liberty of adding up the first 400,000 terms of the Leibniz series using a few lines of computer code[18], and at the end multiplied by 4. This is what his computer program got for π:

$$3.14159...,$$

where after the last digit '9', the digits were no longer correct. Furthermore, this is essentially how the value of π is computed to trillions of decimal places; that is, using a suitable infinite series, as mentioned in Chapter 1. The Leibniz series is not particularly well suited to these lengthy calculations, since it took the sum of 400,000 terms just to achieve a five decimal place accuracy.

A very interesting infinite series is the *harmonic series* given by

$$\frac{1}{1} + \frac{1}{2} + \frac{1}{3} + \frac{1}{4} + \frac{1}{5} + \cdots \tag{4}$$

which acquires its name from the harmonics of music, whereby a vibrating string is divided by nodes into 1, 1/2, 1/3, 1/4, ... equal segments (Fig. 2.5).

When we try to sum the harmonic series, something very curious happens. Clearly the fractions of the sum are becoming very small very quickly. The thousandth number in the series is 1/1000 and the millionth number is 1/1,000,000. In fact, the sum of the first thousand fractions of the harmonic series equals 7.4855 (rounded off to four decimal places). Not a very impressive sum considering we have just added up a thousand terms. How about adding up a million terms? Fortunately, with computers, a simple program can be written to carry out these large sums, but even for a million terms the value of the sum is 14.3927... . We have not even reached a total of 20 yet. But let us carry on this summing process to an even more absurd length, just because we can. This will also demonstrate something important about the nature of infinity.

---

[18] The author used the program Matlab.

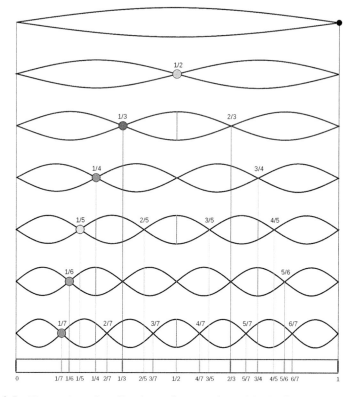

Figure 2.5: Harmonics of a vibrating string, starting with the fundamental tone (1st harmonic) at the top. Then follow the 2nd, 3rd, 4th, 5th, 6th, and 7th harmonics (overtones). The nodes are where the wave displacements cancel each other out. (Image courtesy of Wikicommons.)

Suppose now that our computer can add at the rate of a million fractions each second. In applying our program to the harmonic series, it is found that after ten seconds (that is, ten million summed terms) the computer has now arrived at the princely sum of 16.6953. .. So far, so good, we think to ourselves.

Let us assume we could keep this up, adding another million terms each second, but this time *from the beginning of the Big Bang* until the present. This represents a period of 13.8 billion years, or roughly 10^{17} seconds; that is, 100,000,000,000,000,000 seconds. Now of course, this sum carried on since the Big Bang cannot be carried out in practice, but in the Mathematical Universe it can be estimated that the sum will be less than 55^{19}. Again, not much to show for all those billions of years of addition!

---

[19] With some elementary calculus it is possible to show that the sum of the first n terms of the harmonic series, s_n, is *less than* $1 + \ln n$. Here 'ln' is the natural logarithm (base e) which will be discussed later in the text. In our case $n = 10^6 \times 10^{17} = 10^{23}$, so that $s_n < 1 + \ln 10^{23} = 53.96$, and we have thrown in another whole number for good measure.

Why the insistence on adding up this particular series to more and more terms? The reason is that it can readily be shown that, mathematically, the partial sums s_n of the harmonic series are becoming arbitrarily large, albeit very slowly, as we take more and more terms of the series. Therefore, the sum *in its entirety* is infinite, in the sense that the partial sums s_n will eventually exceed all positive numbers for a sufficiently large value of n[20]. So, we write this simply as,

$$\frac{1}{1} + \frac{1}{2} + \frac{1}{3} + \frac{1}{4} + \frac{1}{5} + \cdots = \infty.$$

Here, we have used the symbol ∞ to mean what it is commonly recognized as; that is, $s_n \to \infty$ as $n \to \infty$. The symbol ∞ itself does not represent a number in any sense, it is purely symbolic. Nor is it a destination that is ever reached.

In other words, the harmonic series is divergent in that it has no finite sum.

We have established that the sum of the harmonic series cannot be finite when summed in its entirety in a mathematical sense, yet when we sum a million terms per second for the age of the Universe (with a capital U), we get less than 55. This extraordinary fact reveals an important distinction between the universe of Mathematics and our own physical Universe. An infinite sum such as the harmonic series can never actually be carried out in the physical world but can effectively be dealt with in the world of Mathematics.

Even more remarkable is the fact that we can remove quite a few terms from the harmonic series and still obtain an infinite sum. In particular, we can just sum the reciprocals of only the *prime* numbers[21],

$$\frac{1}{2} + \frac{1}{3} + \frac{1}{5} + \frac{1}{7} + \frac{1}{11} + \frac{1}{13} + \frac{1}{17} + \frac{1}{19} + \frac{1}{23} + \cdots = \infty.$$

The prime numbers get more and more sparse as we go further and further out in the natural numbers. For example, there are 21 primes between 100 and 200 but only 15 primes between 800 and 900 and only 6 primes between 1,000,000 and 1,000,100. So, the preceding series has fewer and fewer terms, if that is the correct terminology, as there is still an infinite number of terms available to form the infinite sum[22]. Yet the sum remains infinite. In any event, mathematicians are impressed by this and perhaps the reader is as well.

---

[20] As in the preceding footnote, it is possible to prove that the sum of the first n terms of the harmonic series, s_n, are *greater than* $\ln(n + 1)$. Since $\ln(n + 1)$ grows arbitrarily large as $n \to \infty$, then the partial sums s_n must also become arbitrarily large as $n \to \infty$, demonstrating that the harmonic series is divergent. There is also a more elementary demonstration that the harmonic series diverges, without using calculus.

[21] The divergence of this series of the reciprocal of the prime numbers was demonstrated by Euler.

[22] See Euclid's proof in Appendix I that the number of primes is infinite.

Interestingly, if we square the denominator of each term of the harmonic series, we obtain a convergent series whose sum was determined by Leonhard Euler:

$$\frac{\pi^2}{6} = \frac{1}{1^2} + \frac{1}{2^2} + \frac{1}{3^2} + \frac{1}{4^2} + \frac{1}{5^2} + \cdots \tag{5}$$

and we encounter π (and Euler) yet again[23]. Although this result seemingly has nothing to do with circles, one can find a circle lurking in the very distant background.

Leonhard Euler (1707–1783) is one of the true giants of Mathematics, who worked mainly in St Petersburg and Berlin and produced the largest body of mathematical work ever: over 800 publications (including books) in Mathematics, Physics, and Astronomy. He became completely blind during the last 15 years of his life, but in spite of this handicap he still produced hundreds of publications and books thanks to his phenomenal memory and powers of visualization. Euler died while holding a tablet on which he was working out the orbit of the newly discovered planet Uranus. His name will be liberally sprinkled throughout this text.

IN WHAT UNIVERSE IS THIS TRUE?

$$1 + 2 + 3 + 4 + \cdots = -\frac{1}{12}$$

In exactly the same vein as the harmonic series, we find that the series consisting of all the positive integers: $1 + 2 + 3 + 4 + \cdots$ diverges to ∞ as well. Indeed, the partial sums s_n are given by the formula:

$$s_n = 1 + 2 + 3 + 4 + \cdots + n = \frac{n(n+1)}{2},$$

and so it is clear that the partial sums become infinite with increasing values of n, and thus the series diverges. Of course, the reader could see this immediately just by looking at the series.

But, is that the end of the story?

Well, no (see the bottom line in Fig. 2.6 for starters). So, let us pursue this matter a bit further and see where it takes us.

---

[23] This was known as the Bessel Problem and was solved by Euler in 1735.

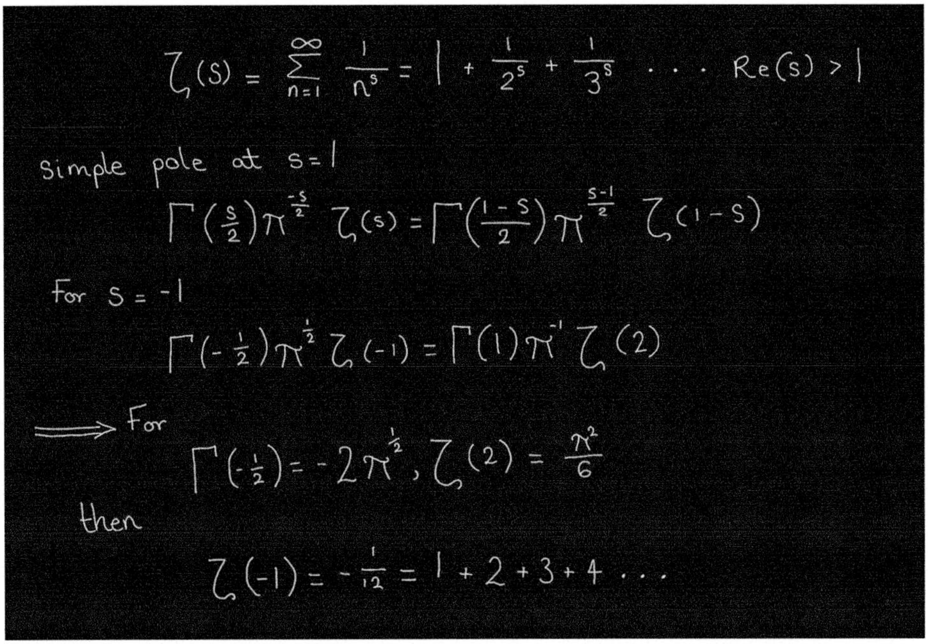

$$\zeta(s) = \sum_{n=1}^{\infty} \frac{1}{n^s} = 1 + \frac{1}{2^s} + \frac{1}{3^s} \cdots \quad Re(s) > 1$$

simple pole at $s = 1$

$$\Gamma\left(\frac{s}{2}\right)\pi^{-\frac{s}{2}} \zeta(s) = \Gamma\left(\frac{1-s}{2}\right)\pi^{\frac{s-1}{2}} \zeta(1-s)$$

For $s = -1$

$$\Gamma\left(-\frac{1}{2}\right)\pi^{\frac{1}{2}} \zeta(-1) = \Gamma(1)\pi^{-1} \zeta(2)$$

$\Longrightarrow$ For

$$\Gamma\left(-\frac{1}{2}\right) = -2\pi^{\frac{1}{2}}, \zeta(2) = \frac{\pi^2}{6}$$

then

$$\zeta(-1) = -\frac{1}{12} = 1 + 2 + 3 + 4 \cdots$$

Figure 2.6: Beauty is in the eye of the beholder. A painting by the author with a 'proof' that $1 + 2 + 3 + 4 + \cdots = -1/12$, using what is called the Riemann zeta function $\zeta(s)$ discussed in Chapter 3. The reasoning is slightly specious at the very last line.

Mathematicians have made a study of divergent series because there is more than one way to look at the 'sum'. In order to do this, let us first recall what a *polynomial* is from high school. It is a mathematical expression that, in its simplest form, involves powers of a single variable x and real number coefficients, such as,

$$x^2 + 2x + 1, \quad x^3 + 3x^2 - 3x + 1, \quad x - 4,$$

and so forth. Put in a value for the variable x and the polynomial determines a numerical value for you. In general, a polynomial in one variable x has the form:

$$a_0 + a_1 x^1 + a_2 x^2 + a_3 x^3 + \cdots a_n x^n,$$

where the coefficients are numbers $a_0, a_1, a_2, \ldots a_n$.

Since we are interested in infinite things in this chapter, what if we let our general polynomial carry on *ad infinitum*? There is really nothing to stop us and, being inquisitive, why not? Suppose we actually do this and give the infinite expression a new name. It is called a *power series*. This name is derived from infinite *series* which we have already encountered and taking *powers* of the variable x. The result will look like:

$$a_0 + a_1 x^1 + a_2 x^2 + a_3 x^3 + a_4 x^4 + \cdots$$

This is just like the preceding high school polynomials, but this one never ends[24]. So, you can think of a power series as just an infinite polynomial, with each a as a fixed constant and the x as some variable, with the juxtaposition of the two meaning they are to be multiplied as for an ordinary polynomial.

Note that by just putting the value $x = 1$ into the above power series, it gives an infinite series of numbers such as the one we started this section with:

$$a_0 + a_1 + a_2 + a_3 + a_4 + \cdots.$$

In fact, putting in any value for the variable x in the power series above will give us an infinite series that then depends solely on the values of x and the coefficients $a_0, a_1, a_2, \ldots, a_n, \ldots$. As we have learned, some infinite series converge nicely and some of them diverge, depending on precisely what the numbers are that we are adding up.

So, with a power series, nothing really new arises here that we need to worry about. We are back on the familiar territory of infinite series.

However, one reason that we consider power series is that many ordinary functions from high school can be written in the form of a power series.

For example, the sine function from trigonometry can be written as:

$$\sin x = \frac{x}{1!} - \frac{x^3}{3!} + \frac{x^5}{5!} - \frac{x^7}{7!} + \cdots, \tag{6}$$

or the cosine function:

$$\cos x = 1 - \frac{x^2}{2!} + \frac{x^4}{4!} - \frac{x^6}{6!} + \cdots, \tag{7}$$

where the value of x is not to be taken in degrees, but in radians. These formulae work for all angles x.

Note that for an acute angle x of a right triangle, the trigonometric function sin x (being the value of the opposite side divided by the hypotenuse) and cos x (being the adjacent side divided by the hypotenuse) are complementary aspects of a right triangle. The above two expressions, (6) and (7), are also complementary, in the sense that sin x involves all the odd powers of x and odd factorials, whereas cos x

---

[24] The power series *will end* if all the constants a_n happen to be zero after some value of n, in which case the power series becomes just an ordinary polynomial. But, in general, we will be considering infinite series that carry on *ad infinitum*.

involves all the even powers of x and even factorials, both of which have alternating signs among the terms.

This is again an example of the 'beauty' found in Mathematics. Just looking at the two infinite series expressions, (6) and (7), anyone would think that the functions they represent must be related in some intimate manner. Indeed, they are.

One virtue of expressions (6) and (7) is that they readily allow one to compute the value of the sine or cosine of any angle. Say for example you wanted to compute the sin $10°$. Then $10° = 0.1745329252$ radians[25] (rounded to 10 decimal places). Putting this value for x into equation (6) and adding up several terms gives the result: sin $10° = 0.173648177$ (rounded to 10 decimal places).

This is exactly the same value given by the author's calculator. He used a three-line Matlab computer program to do this. Indeed, he could effortlessly have added up a few more terms and obtained far more decimal place accuracy than his calculator would have room to display, such is the power of power series[26].

Of particular interest to us now is the function $y = \dfrac{1}{1+x}$, which has a power series representation[27]:

$$\frac{1}{1+x} = 1 - x + x^2 - x^3 + x^4 - x^5 + \cdots$$

with the equality holding for all values of x that are constrained to $-1 < x < 1$. Now, the right-hand side looks very similar to the geometric series discussed earlier. Indeed, the series on the right is the general form of an *alternating geometric series*, where the signs constantly alternate between $+$ and $-$.

What is more, the above expression tells us what the *sum* of such an alternating geometric series will be, namely the value of the left-hand side, $1/(1 + x)$. See Appendix V.

What this means is that by putting in any value of x in the interval between -1 and $+1$, into both the left-hand side and the right-hand side of the expression, both sides will come out equal in value.

To see this power series in action, let us just try a value, say $x = 1/2$, which lies inside our allowable interval of $-1 < x < 1$. Then,

$$\frac{1}{1+1/2} = \frac{2}{3} = 1 - \frac{1}{2} + \frac{1}{4} - \frac{1}{8} + \frac{1}{16} - \frac{1}{32} + \cdots$$

---

[25] Multiplying $10°$ times $\pi/180°$ gives $\dfrac{\pi}{18} = 0.1745329252$ radians.

[26] Actually, hand calculators use a slightly different mathematical method to calculate trigonometric functions, known as the CORDIC (Volder's) algorithm.

[27] Determining the power series representation for this function, as well as the ones given by expressions (6) and (7) and others, is beyond the scope of this book but relatively easy, involving the notion of *Taylor series*.

which is a somewhat interesting result as it is an alternating geometric series. But this not nearly as interesting as what comes next.

Let us now live very dangerously and blithely put in the value $x = 1$ which lies just on the *edge* (and outside) of the allowable interval $-1 < x < 1$, where our alternating geometric series converges to the value given by $1/(1 + x)$. So when $x = 1$, the left-hand side equals the value $1/2$.

Holding our nerve, we put $x = 1$ into the right-hand side of the alternating geometric series and obtain the infinite series of numbers,

$$1 - 1 + 1 - 1 + 1 - 1 + \cdots$$

This infinite series of numbers cannot possibly converge to any (single) finite value in our conventional sense since the partial sums are given by:

$s_1 = 1;$

$s_2 = 0;$

$s_3 = 1;$

$s_4 = 0;$

$\vdots$

and so on, continually oscillating between the values 1 and 0 *ad infinitum.*

The preceding series of the alternating numbers ± 1 is called the *Grandi series*, after the Italian Guido Grandi (1671–1742), and has a long and illustrious history.

So now what? Although the partial sums do not converge to a limit when $x = 1$, what we can do is observe what happens as x *approaches* the value 1 for $-1 < x < 1$, in the alternating geometric series. We will sneak up on the value 1, as it were. Doing so on the left-hand side still gives the value $1/2$ as noted above, but this now provides us with what is called the *Abel sum* of the Grandi series; that is:

$$1 - 1 + 1 - 1 + 1 - 1 + \cdots = \frac{1}{2}.$$

In other words, we can *assign* the value of $1/2$ to the Grandi series by working *outside* the context of conventional infinite series convergence. In another manner of deriving a suitable value for the sum of the Grandi series, there is the *Cesàro sum*, whereby instead of taking the limit of the partial sums as being the sum of the infinite series, one takes the limit of the *averages* of the partial sums (see Appendix VI). The Cesàro sum of the Grandi series is also $1/2$.

Here we have the Abel sum and Cesàro sum of the Grandi series, and it is indeed the case that both sums will also agree with the conventional infinite sum whenever the latter sum exists. But as we have seen with the Grandi series, a conventional sum *does not exist*, and the series is divergent from a conventional point of view.

At this stage, let us return to our original series: $1 + 2 + 3 + 4 + \cdots$, which we know in the conventional sense diverges to ∞; that is, the partial sums become arbitrarily large. Even computing the Cesàro sum for this series gives the same answer.

Undaunted (to be a mathematician you must be fearless), let us give a name to this series to facilitate its manipulation, say:

$$S = 1 + 2 + 3 + 4 + 5 + 6 + \cdots$$

Now we will multiply this series S by 4 (again throwing caution to the wind), with the terms of the resulting series $4S$ placed strategically in line with the terms of S like so:

$$4S = \quad 4 + \quad\ 8 + \quad 12 + \cdots$$

Next let us subtract $S - 4S$ (on both sides) to obtain,

$$-3S = 1 - 2 + 3 - 4 + 5 - 6 + \cdots$$

Nearly done.

Finally, write this same series again, merely shifted over by one digit:

$$-3S = \quad\ 1 - 2 + 3 - 4 + 5 - 6 + \cdots$$

Adding the last two series of $-3S$ and $-3S$ both on the left- and right-hand sides we get the Grandi series again, and we will put in its Abel/Cesàro sum:

$$-6S = 1 - 1 + 1 - 1 + 1 - \cdots = 1/2.$$

All that is left to do now is divide both the left- and right-hand sides by -6 to arrive at: $S = -1/12$, in other words,

$$S = 1 + 2 + 3 + 4 + 5 + 6 + 7 \ldots = -1/12.$$

Of course, the steps we have performed here are not entirely legitimate in the conventional sense, since the series S is actually divergent in our conventional sense and one has to be particularly careful when performing operations involving any infinite series. Nevertheless, in some mysterious unconventional sense, *we can assign a value to* S *of* $-1/12$.

The same result is also obtained by a method known as *Ramanujan summation*. The Indian mathematical genius, Srinivasa Ramanujan, also found this result by the same approach we used above. Furthermore, −1/12 in a related context is the value associated with the famous Riemann zeta function that is discussed in Chapter 3, and of course is the bottom line of the painting in Fig. 2.6.

Srinivasa Ramanujan was a mathematical genius of a sort rarely encountered, in that some of the most abstruse formulas, theorems and conjectures came to him almost as visions. By day, he worked as an accounts clerk in Madras, but at night he worked intensely on Mathematics with little formal training. Thus, his work lacked the sort of rigorous proof that is the gold standard for mathematicians. He wrote to the great English mathematician Godfrey Harold (G.H.) Hardy at Cambridge University in 1913, and Hardy to his credit carried on a correspondence after seeing some of Ramanujan's work. In his second letter, Ramanujan wrote:

> "I have found a friend in you who views my labours sympathetically. This is already some encouragement to me to proceed... I find in many a place in your letter rigorous proofs are required and you ask me to communicate the methods of proof.... The sum of an infinite number of terms of the series $1 + 2 + 3 + 4 + \cdots = -1/12$ [is true] under my theory. If I tell you this you will at once point out to me the lunatic asylum as my goal.... What I tell you is this. Verify the results I give and if they agree with your results, got by treading on the groove in which the present-day mathematicians move, you should at least grant that there may be some truths in my fundamental basis[28]."

Furthermore, the sum, $1+2+3+4+\cdots = -\dfrac{1}{12}$, does make an appearance in String Theory[29], one of the foremost attempts to explain the workings of our Universe at its smallest scale. In the words of theoretical physicist, Lee Smolin, "If String Theory is a mistake, it's not a trivial mistake. It's a deep mistake and therefore kind of worthy."

Our ridiculous sum also comes up in the study of the Casimir Effect, which is the very small attractive force found between two close (within nanometers) parallel plates in a vacuum due to quantum fluctuations in the region surrounding the plates.

---

[28] Ramanujan joined Hardy and John Edensor (J.E.) Littlewood at Cambridge in 1914 and a plethora of wondrous Mathematics resulted, much of it over the ensuing decades after he died in 1920 at the age of 32. An excellent film, *The Man Who Knew Infinity* (2015) has been made of his life.

[29] Joseph Polchinski, *String Theory*, Cambridge University Press, 1998, p. 43.

THE POWER OF *e*

There is another pre-eminent number like that of π which is not as well known to the general public, but is perhaps equally, if not more important in the world of Science and Mathematics. The number is denoted by the letter *e*, which is in honor of Euler. One way to define the number *e* is by the infinite series

$$e = 1 + \frac{1}{1!} + \frac{1}{2!} + \frac{1}{3!} + \cdots + \frac{1}{n!} + \cdots \tag{8}$$

where $n! = n(n-1)(n-2)\cdots 2 \cdot 1$, is '*n* factorial'. The series sums in its entirety to the value $e = 2.718281828459045\ldots$

While *e* is just a number, albeit an important one, we are also interested in taking arbitrary powers of this number, which leads to the *exponential function* e^x given by the elegant power series expression

$$e^x = 1 + \frac{x}{1!} + \frac{x^2}{2!} + \frac{x^3}{3!} + \cdots + \frac{x^n}{n!} + \ldots, \tag{9}$$

which is valid for any finite number *x* whatsoever.

You can see that by putting in the value $x = 1$ into eq. (9) you get eq. (8). But the virtue of eq. (9) is that we can put in any value we wish for *x* and the power series will churn out the value of e^x.

This leads us to why eq. (9) is so cool. We could put in say the value $x = 2$ into both sides of the equation, add up a goodly number of terms on the right-hand side and get a numerical value for e^2. But doing this would be a complete waste of time as we could just multiply $e \times e$ to get the value of e^2. As Homer Simpson would say, "doh".

But now what if we want to put in $x = \pi$ in order to compute e^π? What does this expression mean? Does such an expression even make sense? Well yes, it does, and in Chapter 3 we will want to consider just such powers of *e*. We certainly cannot calculate e^π by multiplying *e* pi times itself, which makes no sense whatsoever.

But what we can do is put a value of $x = \pi$ (rounded off to any number of decimal places) into eq. (9), preferably on a computer and, taking the sum of a reasonable number of terms, we will arrive at a value of e^π.

Just in case you were wondering, the author wrote a little program to compute e^π. It came out to: $e^\pi = 23.13$. Likewise, we could compute e^e or $e^{\sqrt{2}}$ since eq. (9) works for any value of *x* you can think of. Take that Homer.

Strange as it may seem, the number *e*, as well as the function e^x, occurs in many aspects of the real world, as we shall see.

Actually, there are various ways to define e, as well you might expect from such a significant value intrinsic to the natural world. All the definitions are equivalent to one another.

Here is another way that provides some insight into computing compound interest. First, consider the quantity

$$\left(1 + \frac{1}{n}\right)^n,$$

for a few numbers n that get larger and larger. Say for example, $n = 1, 2, 3, 1000$, $1,000,000$ and lastly, $1,000,000,000$. At first glance, we notice that the quantity inside the parenthesis is getting smaller, but on the other hand the exponent n is simultaneously getting larger. So, it is not yet really obvious what is happening with $(1 + 1/n)^n$ for ever increasing values of n.

But it becomes much more obvious if we tabulate our results:

Table 2.1. Results of computing $(1 + 1/n)^n$ taking increasing values of n. The results converge to the value of e.

n	$(1 + 1/n)^n$
1	2
2	2.25
3	2.3703703...
1,000	2.7169239...
1,000,000	2.7182804...
1,000,000,000	2.7182818...

As you can see from Table 2.1, the values of the quantity $(1 + 1/n)^n$ being calculated are not becoming arbitrarily large or even behaving in an erratic manner. In fact, values are converging to a specific value that has, as its first five decimal places: 2.71828.

If we should choose to take ever larger values for the number n, we would get ever more digits of the constant $e = 2.718281828459045...$. In our mathematical parlance, the number e represents the *limit* of this never-ending process of calculating the value of the quantity in question, for higher and higher values of the number n. We can write this limiting process in the slick mathematical form:

$$\left(1+\frac{1}{n}\right)^n \to e = 2.71828182845\ldots, \tag{10}$$

as $n \to \infty$[30].

In plain English, eq. (10) is the mathematical way of saying that the quantity $(1 + 1/n)^n$ is approaching the value e as the number n gets larger and larger. The symbol ∞ only denotes the fact that the number n is to increase indefinitely in value. It never arrives at a destination since there is none.

Although, the quantity $(1 + 1/n)^n$ never actually arrives at its destination, it does at least have one that it approaches in the limit, namely, e. The never-ending string of decimal values of e only gradually reveal themselves to us as we take larger and larger numbers n. Like π, it can never be known to us in its entirety, but in 2016 the first five trillion digits of e were calculated by Ron Watkins using over 48 days of computer time. Beyond that lies an infinite sea of more digits, some of which may never be discovered, but in some ethereal sense the digits are there, waiting.

Given the infinite series definition of e in eq. (8) and summing up the terms on a computer for sufficiently large values of n, we would obtain exactly the same numerical value for e as in computing the value of $(1 + 1/n)^n$ for ever larger values of n.

A slight variation of the above is if we replace the fraction $1/n$ by x/n in eq. (10), which yields the formerly defined exponential function given by eq. (9):

$$\left(1+\frac{x}{n}\right)^n \to e^x,$$

as $n \to \infty$. This provides another way to consider the function e^x.

Note that if we let $x = -1$ in this formulation then we find that,

$$\left(1-\frac{1}{n}\right)^n \to e^{-1} = \frac{1}{e} = 0.367879\ldots,$$

as $n \to \infty$, which is an important number to keep in mind when gambling.

Suppose that you go to Las Vegas and place a bet on a single number at the roulette table. The odds of that number coming up on the roulette wheel are one in 38[31].

---

[30] This limiting process is already familiar to us as it is the same notion involved in determining the limit of the partial sums s_n for some infinite series, as $n \to \infty$.

[31] A roulette wheel in the U.S. has numbers 1 through 36 plus 0 and 00. European wheels lack the latter 00, so your odds in Europe are one in 37 when betting on a single number.

So, you will most likely lose on the first spin, but think to yourself that your number has the same odds as any other, so you leave your money on it. Surely you think, you will win at least once in 38 spins, and hopefully well before then. Sorry to say, your probability of losing *all* 38 times in a row is not negligible[32]:

$$\left(1-\frac{1}{38}\right)^{38} = 0.363 \approx \frac{1}{e},$$

which, as we just saw above for large values of n, will be a good approximation for the value of $1/e$. Hence on average, this run of rather bad luck can happen about one-third of the time. The same principle applies in trying to toss a die six times in order to get the number 3. Your odds of complete failure are much the same as for the faulty roulette strategy. You have been warned.

Interestingly, this value of $1/e$ turns up in another unusual situation commonly known as the 'secretary problem' or 'marriage problem', of which there are several variations. Without getting too specific, suppose that one wishes to choose a secretary for a particular job among a large number, say N, of applicants. Each applicant is interviewed one after the other and given a ranking against all the previously interviewed applicants. The catch here is that it must be decided on the spot whether or not to hire the current applicant, based on their ranking with respect to the previous applicants. Naturally, of course, you only want the best.

For a very large value of N, say 500 applicants, one could well imagine that an optimal strategy would be desirable instead of having to interview every last applicant. But is there an optimal strategy to determine how many applicants to interview? There is, but of course it will not always turn up the very best candidate.

It turns out that the best strategy is to interview the first $1/e$ applicants (about 37% of the total number N) and then select the very next applicant who out ranks all those previously interviewed. Furthermore, it turns out that the probability of the very best applicant being selected by this method is also at least $1/e$[33].

Famed astronomer Johannes Kepler (1571−1630), with whom we will have a serious encounter in Chapter 6, faced a somewhat similar situation regarding his choice of a new wife after his first wife died. Over almost two years of intensive investigation, involving interviewing and assessing the relative merits of no fewer than 11 women for the position of wife, Kepler chose number five on his list.

---

[32] Since the probability of winning at each spin of the roulette wheel is 1/38, the probability of *losing* on each spin is $\left(1-\frac{1}{38}\right)$. Therefore, losing 38 times in a row is given by $\left(1-\frac{1}{38}\right)^{38}$.

[33] It should be noted that the value of $1/e$ in each instance is only approached in the limit as n increases. So, the value is only ~$1/e$ in each case.

Note that 37% of 11 is 4 (rounded off) and the very next person on the list would be number 5. In the words of statistician Thomas Ferguson, "Perhaps, if Kepler had been aware of the theory of the secretary problem, he could have saved himself a lot of time and trouble[34]."

FAST MONEY

Here is a mundane application that is guaranteed to make money. Perhaps not as much as you would think, but it does happen to be directly related to the exponential function. The standard formula for compound interest is:

$$A = P\left(1+\frac{r}{n}\right)^{nt},$$

where A is the final amount of principal and interest, P is the original principal, r the interest rate, n the number of times the interest is compounded per year, and t is the number of years over which the interest accrues.

The expression in the parenthesis of this formula has a familiar ring to it when compared to the ones above, and in fact, if we let the number of times per year that the interest is compounded become arbitrarily large, as we did above in letting $n \to \infty$, then we find that the amount A is approaching the value Pe^{rt} which is the value of your earnings when the bank says that your interest is 'compounded continuously'; that is[35],

$$A = Pe^{rt}.$$

This makes it possible for a computer to determine your earnings at the end of a given time period, t, and interest rate, r, without having to calculate it at every moment of every day, as the term 'compounded continuously' might suggest. It is through this type of calculation of compound interest that the number e was essentially discovered by Jacob Bernoulli 1655–1705) when studying this very issue. However, others played an indirect role when dealing with the notion of logarithms[36].

Yes, this is rather pedestrian compared to the Theory of Relativity, but it does clear up one of life's little mysteries about how a bank can compound your interest continuously.

In various cases to come, we will be interested in the values of the reciprocal of the exponential function: $1/e^t = e^{-t}$, for positive values of t (time). This function now decreases, and in a particular way that will be essential in describing many natural processes (Fig. 2.7).

---

[34] T. Ferguson, Who Solved the Secretary Problem? *Stat. Sci.*, Vol. 4, No. 3, pp. 282–289, 1989.

[35] $\left(1+\frac{r}{n}\right)^n \to e^r$ as $n \to \infty$ as above, replacing x by r.

[36] The logarithm function is the inverse of the exponential function, so they are closely related. Thus, the number e occurs implicitly in earlier work regarding logarithms by John Napier and others.

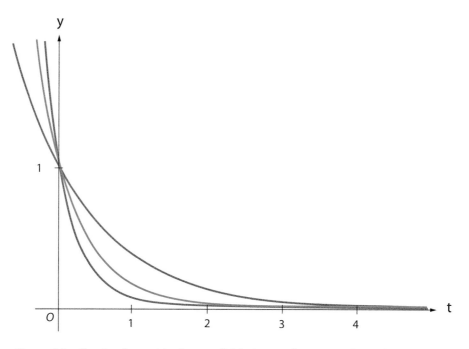

Figure 2.7: Graphs of $y = e^{-t}$ (red), $y = e^{-2t}$ (blue), $y = e^{-3t}$ (green). (Illustration courtesy of Katy Metcalf.)

WHAT IS NORMAL?

One very important application of the reciprocal of the exponential function is in the field of statistics. If one collects data on, say, the systolic (the number on top) blood pressure of a large group of adults, and set out all the blood pressure values along the x-axis and their corresponding *frequency of occurrence* (i.e. the fraction of the total) along the y-axis, the result is a bell-shaped curve centered about the mean (average) value, as in Fig. 2.8.

For the sake of simplicity, if we take the mean to be zero, then our curve will have the form of the basic equation:

$$y = \frac{1}{\sqrt{2\pi}} e^{-x^2/2}.$$

It should be noted that this curve is an idealization, since in the equation as $x \to \infty$, then $y \to 0$, but in reality, systolic blood pressure cannot become infinite as this soon leads to death.

In the real world, of course, the mean is not necessarily zero either (again like systolic blood pressure), and the curve can be narrow and tall with the majority of values not very far dispersed away from the mean, or short with widely dispersed values. The measure of the amount of dispersion about the mean is the *standard deviation* and is commonly denoted by the symbol σ[37].

So, to accommodate a particular mean value which we denote by μ, and a standard deviation σ denoting the amount of dispersion of the values about the mean, we now have the more general formula for the frequency of occurrence:

$$y = \frac{1}{\sigma\sqrt{2\pi}} e^{-\left(\frac{x-\mu}{\sigma}\right)^2 / 2},$$

which is the basic formulation for a 'normal distribution' in statistics (Fig. 2.8). It looks a bit scary, but we are not going to use this formula as it is just here for show, but the notion of the standard deviation σ is a very important one. Among countless other things, it plays a central role in the equation for the measurement of the mass of a supermassive black hole, as will be seen in Chapter 6.

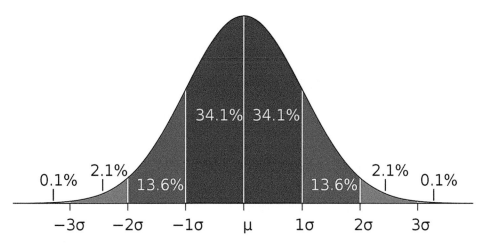

Figure 2.8: The normal distribution curve (bell curve) which is described by an exponential function. The mean occurs at $x = \mu$, with the curve completely symmetric about the mean. The image shows that 68% of the data points are found within one standard deviation (1 σ) either side of the mean, 95% of the data points are within two standard deviations, 99.7% of all the data fall within three standard deviations, and so forth (values have been rounded off). (Image in public domain.)

---

[37] The standard deviation can be given a rigorous definition, as found in any book on statistics, but the general idea suffices here.

Often in Science, an occurrence is referred to as a '5-sigma event'. This means that the frequency of its occurrence would be that of a data point which lies 5-sigma (five standard deviations) from the mean in a normal distribution. This works out to about one chance in 1.75 million, taking both sides of the distribution into account, or one in 3.5 million considering only a single side.

It is also used as the gold standard for discovering something new in Physics. The discovery of a new particle, such as the Higgs boson, must be at the (one-sided) 5-sigma level; that is, the likelihood that the experimental signal indicating the Higgs particle was simply a random background fluctuation would be no more than one in 3.5 million.

In other sciences, often a 2-sigma or 3-sigma result is sufficient, which means that the likelihood of the result being due to chance alone is less than 5% (1 in 20) or 0.3% (1 in 300) respectively[38].

MULTIPLYING *AD INFINITUM*

Not only is there a way to add up an infinite collection of numbers, but it is also possible to multiply an infinite set of numbers, such as, a_1, a_2, a_3, a_4..., employing a similar technique as we did for the addition of an infinite set of numbers. Instead of looking at the partial sums, we consider the *partial products*:

The first partial product is : $p_1 = a_1$;

The product of the first two numbers is : $p_2 = a_1 \cdot a_2$;

The product of the first three numbers is : $p_3 = a_1 \cdot a_2 \cdot a_3$;

$\vdots$

So that the *nth* partial product is the product of the first n numbers:

$$p_n = a_1 \cdot a_2 \cdot a_3 \cdots a_n.$$

Just as we did with partial sums, we look for what value (limit) if any, the partial products p_n are approaching closer and closer to as we take more and more terms; that is, as n increases indefinitely. If there is such a value, say P, and it is not zero, then we say that P is the value of the infinite product and write,

$$P = a_1 \cdot a_2 \cdot a_3 \cdot a_4 \cdot a_5 \cdot a_6 \cdots a_n \cdots$$

_____

[38] Recently, these levels of significance have come under some scrutiny: 'Scientists rise up against statistical significance', Comment, *Nature*, March 20, 2019.

Once again, our number π is found lurking in a beautiful infinite product expression with doubly repeating even and odd numbers:

$$\frac{\pi}{2} = \frac{2 \cdot 2 \cdot 4 \cdot 4 \cdot 6 \cdot 6 \cdot 8 \cdots}{1 \cdot 3 \cdot 3 \cdot 5 \cdot 5 \cdot 7 \cdot 7 \cdots},$$

and once again the equality is exact. This formula is attributed to English mathematician and clergyman John Wallis in 1655, and many proofs of it have been given over the centuries. Rather remarkably, this very formula is a consequence of quantum mechanical considerations of energy levels of the hydrogen atom discovered just recently[39].

So here we have an interesting formula from the world of pure Mathematics that can be derived from arcane formulations in Quantum Mechanics (a subject discussed in a later chapter). It is hoped that the reader finds this as truly astonishing as the author does.

---

[39] T. Friedmann, C.R. Hagen, Quantum mechanical derivation of the Wallis Formula for π, *J. Math. Phys.*, 56, 112101, 2015.

3

Imaginary Worlds

The imaginary number is a fine and wonderful resource of the human spirit,
almost an amphibian between being and not being...
Gottfried Wilhelm Leibniz (1646−1716)

God made the universe out of complex numbers...
Mathematician Richard Hamming[1]

Perhaps the most bizarre creature in the mathematical menagerie is that of the imaginary number, *i*. The adjective 'imaginary' is somewhat unfortunate, because to an engineer, physicist or mathematician it is something very tangible, though the layperson would not recognize it as a number at all. It is nowhere to be found on the number line and does not really exist in Euclidean two-dimensional space either. So where does it come from and where does it exist?

As the author works in the field of complex numbers, he concurs with Hamming's assessment.[1]

---

[1] The full quote is: "A few years ago I had the pleasure of teaching a course in complex variables. As always happens when I become involved in the topic, I again came away with the feeling that, 'God made the universe out of complex numbers'." From his article, The Unreasonable Effectiveness of Mathematics, *The American Math. Monthly*, Vol. 87 No. 2, pp. 81–90, 1980.

© Springer Nature Switzerland AG 2020
J. L. Schiff, *The Mathematical Universe*, Springer Praxis Books,
https://doi.org/10.1007/978-3-030-50649-0_3

THE STRANGE CASE OF $x^2 + 1 = 0$

It has been known since Babylonian times that if you multiply any two positive numbers together then the answer is positive. It was only in the 7th century that mathematician and astronomer Brahmagupta published that the product of any two negative numbers was also positive. Therefore, any real number, whether positive or negative, multiplied by *itself*, will be positive. In other words, there is no real number multiplied by itself that will ever give a negative number for an answer.

Stating this in yet another way, there is *no real number solution* to the expression,

$$x^2 = -1,$$

where x represents any number. This was a matter exercising the mind of the outstanding Indian mathematician Bhaskara back in the 12th century, who proclaimed that this equation was impossible to solve. By adding 1 to both sides of this equation, it takes on an equivalent form: $x^2 + 1 = 0$, which is a 'quadratic equation[2]' of the sort most of us have encountered in high school. But still there is no real number solution and in this sense, Bhaskara was correct.

So, is the above equation untouchable? If you are tempted to take the square root of both sides of the equation and write:

$$x = \pm\sqrt{-1}$$

you are on the right track, but as discussed above, the value of x is not any real number and these are the only kind of numbers that we know of thus far.

This was the state of affairs until the 16th century, when the Italian physician and scientist Gerolomo Cardano found a way out of this dilemma in his major study of techniques for solving algebraic equations, *Ars Magna* (1545). Cardano's solution when encountering the square root of negative numbers was simply to appropriate them and not dismiss them from his calculations[3]. This made Cardano uneasy, calling the new numbers, "…the closest to the quantity that is truly imaginary…".

---

[2] In a quadratic equation, the greatest power of the variable x is 2. Cubic equations have 3 as the greatest power of x, etc.

[3] Cardano's problem was to find two numbers whose sum was 10 and product was 40. Letting x and $10 - x$ represent the two numbers, multiplying them leads to the quadratic equation,

$$x^2 - 10x + 40 = 0.$$

In modern parlance, for those who remember the quadratic formula, this leads to the two solutions

$$x_1 = 5 + \sqrt{-15} \text{ and } x_2 = 5 - \sqrt{-15}.$$

Fortunately, when x_1 and x_2 are added, the troublesome $\sqrt{-15}$ disappears and the sum is 10. Likewise, the product has the value $25 - (-15) = 40$, as the cross terms cancel out. This says that x_1 and x_2 are indeed solutions to Cardano's problem and you just have to live with the outcome.

Some years later, the Italian mathematician Rafael Bombelli described how to add, subtract, and multiply numbers involving square roots of negative numbers, in his seminal treatise *L'Algebra* (1572), and the theory of such numbers was underway.

But rather than explaining Bombelli's approach, which is one still in use today, let us take a parallel approach introduced by a famous Irish mathematician, William Rowan Hamilton, in the 19th century. Instead of considering numbers on the number line, let us consider instead points in the two-dimensional (*Euclidean*) plane as pairs of real numbers[4], as in Fig. 3.1.

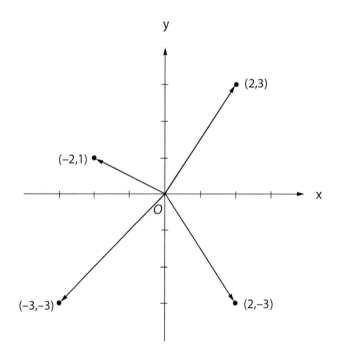

Figure 3.1: Some ordered pairs of points in the two-dimensional Euclidean plane. The same points will be viewed from a slightly different perspective in Fig. 3.2. (Illustration courtesy of Katy Metcalf.)

Now every point in the plane has a unique two-coordinate designation known as an *ordered pair*, (a,b), with the horizontal coordinate always stipulated first by convention. This system gives a location to every point in the plane, much like every point on Earth can be located by its longitude and latitude coordinates.

All the numbers along the horizontal axis can also be designated as either a or $(a,0)$, since they both represent the same point. Similarly, all the points on the vertical axis can be designated as either b or $(0,b)$.

---

[4]The ordered pairs of numbers depicted in Fig. 3.1 all involve integers (just to make their location more obvious) but the two-dimensional Euclidean plane consists of all ordered pairs of real numbers.

The great leap forward (made by Bombelli in his parallel context and invoked by Hamilton), was to treat any *pair*,

$$(a, b)$$

in the plane, as a *new kind of number* in its own right. But if you are going to create a new class of numbers, then you had better explain how to add, subtract, multiply, and divide these new numbers, since that is what one does with numbers.

Addition is pretty obvious, and if you were to make up your own natural way of adding ordered pairs of real numbers you would probably come up with the following definition:

$$(a, b) + (c, d) = (a + c, b + d),$$

where the addition on the right-hand side is simply the ordinary addition of real numbers. Therefore, the sum of the first two coordinates gives the first coordinate of the sum, and the sum of the second two coordinates gives the second coordinate of the sum. What could be nicer.

It is clear that adding two of the new kinds of numbers gives an answer of exactly the same type – that is, an ordered pair – which is as it should be. When we add two real numbers together, we expect the answer to be a real number as well, not a house or a dog.

Subtraction is done likewise, but multiplication and division[5] are a bit more arcane. Here, we will only consider multiplication, which is given by the rule:

$$(a, b) \times (c, d) = (ac - bd, ad + bc).$$

However, a very important consequence from the rule for multiplication is that,

$$(0, 1) \times (0, 1) = (-1, 0).$$

Now as we mentioned above, a pair of the form $(a, 0)$ is just another way to express the real number a, which means that the number $(-1, 0)$ on the right-hand side of the equation is the same as -1. Thus, in terms of our new numbers and our new rule for multiplying them,

$$(0, 1) \times (0, 1) = (-1, 0) = -1.$$

We have now found a new kind of number, represented by an ordered pair (a, b), combined with a few rules of how to do arithmetic with them, and we discovered that when we multiply $(0, 1)$ times itself, we obtain the answer -1. How about that?

---

[5] Division is a variation of the rule for multiplication and is not necessary for our purposes.

THE 'i's HAVE IT

Because of the very singular property of $(0, 1)$, that when it is multiplied by itself it returns the value -1, we give $(0, 1)$ a specially designated symbol, i. Thus i has the property that:

$$i^2 = i \times i = -1,$$

which is certainly *not* the case for any real number[6]. So i is a so-called 'imaginary number', but as you can see there is really nothing imaginary about it whatsoever. It is simply the ordered pair $(0, 1)$ together with the appropriate rule for multiplication.

This last equation allows us to write,

$$i = \sqrt{-1},$$

which, it has to be admitted, does lend the imaginary number a certain aura of mystery because it is definitely not a real number. This approach via ordered pairs of real numbers with the above rules for doing arithmetic goes back to William Rowan Hamilton, as mentioned above, and whom we shall encounter again in due course.

Let us consider all ordered pairs of real numbers (a, b), which are just all the points in the two-dimensional plane as in Fig. 3.1. Now *couple* these points with the appropriate rules of addition, subtraction, multiplication, and division. The result is a new number system with which we can do arithmetic, just like with ordinary real numbers. This new system of numbers has a name: they are called *complex numbers*. But they are not really complex in the sense of 'complicated', they are just different from real numbers in that they have two components and their arithmetic abides by their own set of rules. Different strokes for different folks.

You may very well ask, at this juncture, how does one distinguish a complex number (a, b) from the ordered pair (a, b) in the Euclidean two-dimensional plane, without mentioning the accompanying rules for addition, subtraction, multiplication, and division? In order to avoid any confusion, we put the *complex number* (a, b) into a new polynomial-like format:

$$a + bi.$$

This new designation tells us that we are dealing with a complex number and not just a point located at the position (a, b) in the plane[7]. The complex number comes with its own baggage, so to speak, namely specific rules of arithmetic for dealing with it.

---

[6] It should be noted that the number $-i$ has the same property, that is, $(-i)^2 = -1$, but there are no other such ordered pairs whose square equals -1.

[7] One can write a complex number as $a + bi$ or $a + ib$. They represent exactly the same number and the use of either form is just a matter of personal preference. We will sometimes use both formats for aesthetic reasons.

For any complex number $z = a + bi$, the two components a and b have names for easy reference. The number a is called the *real part* and the number b is called the *imaginary part*. Of course, both a and b are real numbers, but the number b is attached with the imaginary number i, hence its name.

Moreover, the bi part of $a + bi$ really is $b \times i$, such as $3i$, $-3i$, as indicated in Fig. 3.2. Therefore, any real number a can be considered a complex number of the form:

$$a = a + 0i,$$

and this means that the set of real numbers is a subset of the set of complex numbers.

The rules that we stipulated for the arithmetic of ordered pairs, but with our new notation for complex numbers, again tells us how to do arithmetic with them.

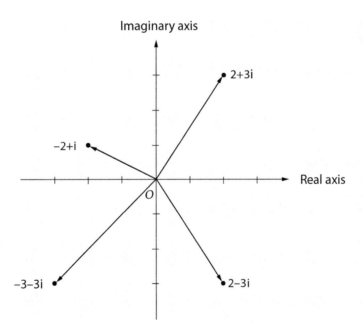

Figure 3.2: The same set of points as in Fig. 3.1 but here considered as complex numbers in the environment of the complex plane, with its own rules for arithmetic. (Illustration courtesy of Katy Metcalf)

The collection (set) of all the complex numbers, $a + bi$, or if you prefer, $a + ib$, for all real numbers a and b, form the set of complex numbers, $\mathbb{C}$. In this new number system, we can add, subtract, multiply and divide. The complex numbers behave very nicely, much like our real numbers (denoted by $\mathbb{R}$), in that the numbers in $\mathbb{C}$ are commutative and associative for both addition and multiplication. But they do lack one property of the real numbers, in that they cannot be ordered in the sense that one number is larger or smaller than another.

The reason for the great mystery surrounding the imaginary number, i, is because of its special nature. This simple act of creation of a new type of number system, the complex numbers, $\mathbb{C}$, together with the accompanying four rules of combining them: $+, -, \times, \div$, yields an entirely new, profound, and richly beautiful branch of Mathematics, called Complex Analysis.

Surprisingly, this new family of numbers has become absolutely fundamental to our understanding of physical reality. Complex numbers are basic to the study of electrical circuits, in the study of how fluids flow, and of signal processing. They are at the heart of the Theory of Relativity, and a ubiquitous feature of Quantum Mechanics, the study of the nature of our Universe in its smallest aspects.

It is fair to say that complex numbers are as much a part of understanding the Universe as are the real numbers that we are all familiar with.

Because complex analysis has this deep and profound aspect to it, one can almost certainly believe that it was just waiting in the netherworld of Mathematics for Cardano and Bombelli to come along and snatch it into the real world. Without a doubt, both gentlemen would have been astounded to know just how important their discovery would turn out to be.

A more complete history of the imaginary number $\sqrt{-1}$ can be found in Paul Nahin's fine account: *An Imaginary Tale: The Story of $\sqrt{-1}$* (see Bibliography).

THE GOD-LIKE EULER IDENTITY

In order to obtain a very profound result regarding complex numbers and the Mathematical Universe, we just need to consider a simple bit of trigonometry.

As per Fig. 3.2, we can depict the same ordered pairs seen in Fig. 3.1, but now in their new complex number format. If we draw a straight line from the origin O to the point $z = x + iy$ in the complex plane environment, as in Figs. 3.1 and 3.2, that line segment is called a *vector*, having a length and direction. Both its length and direction are of particular significance. Firstly, we give a name to the length of the vector, which we call the *modulus* of z and denote it by $|z|$[8].

The modulus of a complex number − actually the square of the modulus − plays an extremely important role in Quantum Mechanics, as will be discussed in Chapter 7 and is depicted in Fig. 7.3 therein. It is related to the manner in which one can describe the location of a subatomic particle.

Considering again the point $z = x + iy$ in the complex plane, from basic trigonometry (Fig. 3.3) we have:

$$x = r\cos\theta, \; y = r\sin\theta,$$

---

[8] By the Pythagorean theorem, the length $r = |z|$ is given by: $r = \sqrt{x^2 + y^2}$, which is the modulus (or *absolute value*) of z. If we instead consider the point z as a vector in the Euclidean two-dimensional plane determined by coordinates (x, y), then $|z|$ denotes the length of that vector too. So, the concepts match up in both the Euclidean and complex planes.

where θ represents the *argument* of z, which is just the angle (in radians) from the x-axis to the vector formed by z, and $r = |z|$. Like the modulus, the argument is a special quantity worthy of a name. The reason for the significance of r and θ is that we can also express the complex number, $z = x + iy$ solely in terms of r and θ in the form:

$$z = r\cos\theta + ir\sin\theta = r(\cos\theta + i\sin\theta),$$

which is often very convenient as we will see shortly.

The geometric description of complex numbers lying in a plane composed of a real (horizontal) axis and an imaginary (vertical) axis, as in Figs. 3.2 and 3.3, was first described by a Norwegian-Danish surveyor, Caspar Wessel (1745–1818), and subsequently independently by Swiss amateur mathematician Jean-Robert Argand (1768–1822). Sometimes, the plot of complex numbers in the plane is referred to as an 'argand diagram' although Wessel's description pre-dates Argand's by some nine years.

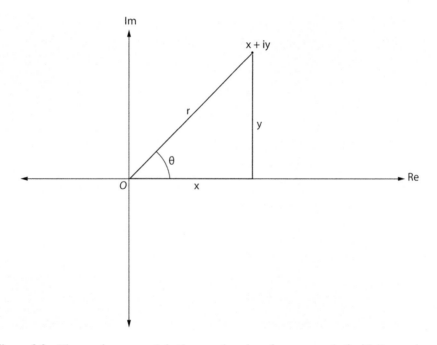

Figure 3.3: The *number* $z = x + iy$ in the complex plane forms an angle θ with the x-axis so that $x = r\cos\theta$, $y = r\sin\theta$. It follows that: $z = r(\cos\theta + i\sin\theta)$. (Illustration courtesy of Katy Metcalf.)

Indeed, let us consider the set of all the points $z = r(\cos\theta + i\sin\theta)$, for which their modulus $r = |z| = 1$. Now what exactly is this set of points z whose distance from the origin O is 1? The answer, which should be obvious, is simply a circle of radius $r = 1$ in the complex plane. Moreover, its equation is given by:

$$z = \cos\theta + i\sin\theta,$$

for $0 \leq \theta \leq 2\pi$. This circle is called the *unit circle* because its radius is one unit[9].

Now for a bit of mathematical fireworks. Some 250 years or so ago, the great Swiss mathematician Leonhard Euler took the previous series representations for the sin x (eq. 6) and cos x (eq. 7), added the two series together (we will replace x by θ as is customary) and formed the complex number with the two series:

$$\cos\theta + i\sin\theta.$$

Remarkably, for Euler and the rest of the world, the sum, upon simplification, turned out to be the series representation for the exponential function $e^{i\theta}$, given by eq. (9), with $x = i\theta$. That is, we have arrived at the famous 'Euler formula',

$$e^{i\theta} = \cos\theta + i\sin\theta, \tag{11}$$

where again θ is in radians (see Fig. 3.4).

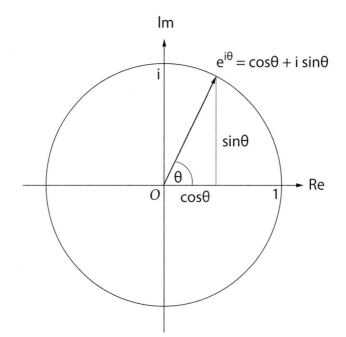

Figure 3.4: The set of points in the complex plane that are 1 unit from the origin form a circle of radius 1. Every point on this unit circle can be represented as: $z = \cos\theta + i\sin\theta = e^{i\theta}$ according to Euler's formula. (Illustration courtesy of Katy Metcalf.)

---

[9] In the ordinary two-dimensional plane, all points (x, y) on a circle of radius 1 satisfy the equation: $x^2 + y^2 = 1$.

Intuitively, without going into the details of the calculation, the power series for cos θ (eq. 7) involves all the *even* powers of θ divided by even factorials. At the same time, the power series for sin θ (eq. 6) involves all the *odd* powers of θ divided by odd factorials. On the other hand, the series representation for e^θ (eq. 9) involves *all* the powers of θ divided by their respective factorials. This suggests that by adding the series for cos θ and sin θ, the sum should resemble the series for e^θ.

Now specifically, if we add the two series for cos θ + i sin θ, and compare with the series representation for $e^{i\theta}$, all the pluses, minuses and imaginary numbers are identical on each side of eq. (11). The Euler formula once again.

If Physics has the Einstein equation, $E = mc^2$, showing that energy (E) and mass (m) are really just different forms of the same thing, connected by the constant c, the speed of light, then Mathematics has the following equation, obtained by setting $\theta = \pi$ (radians) in Euler's formula (eq. 11)[10]. Behold, the *Euler identity*

$$e^{i\pi} + 1 = 0. \tag{12}$$

To many, including this author, this formula is almost supernatural in its appearance[11].

Here, we are raising the number e to the power $i \times \pi$, which in itself is rather mysterious. Can we even raise a real number like e to a *complex* power like $i\pi$? It turns out we can but we will not go into the details here. In any event, we know from Euler's formula that its value is given by: $e^{i\pi} = \cos \pi + i \sin \pi = -1$.

Moreover, the numbers e and π are both infinite decimal expansions. Somehow, by throwing into the mix the imaginary number i and then adding 1, everything disappears as if by magic, and we get nothing, that is to say, 0. There is something divine, even mystical about the Euler identity. It just seems too astounding that e, i, π, 1, and 0, can be related so simply in a single meaningful equation.

Although it is not obvious how to do this, it is even possible to define a complex number raised to the power of another complex number. One interesting consequence is the only slightly less mystical result of raising the imaginary number i to the power i,

$$i^i = e^{-\pi/2} = 0.20788\ldots$$

again involving i, e, and π, with the result being a *real* number[12].

Perhaps this is what 18th century German philosopher Georg Friedrich Philipp von Hardenberg had in mind when he said that, "Pure mathematics is religion."

---

[10] Note that π radians = 180° and that cos π = -1 and sin π = 0.

[11] Indeed, some have used eq. (12) to assert the existence of God. See Pickover 2009 reference in the Bibliography, p. 232. Celebrated physicist Richard Feynman called it "the most remarkable formula in mathematics."

[12] This is known as the 'principal value' of i^i. Raising a complex number to the power of another complex number involves the natural logarithm which has a multiplicity of values, but one can be distinguished as the principle value, as in the text.

EVEN MORE IMAGINARIES – QUATERNIONS

Since we have extended our real number system by creating the complex numbers, can we do likewise and extend the complex numbers further? Indeed, why stop with just adding one imaginary component to a real number as we did for complex numbers: $z = x + yi$?

In fact, the Irish mathematician William Rowan Hamilton did try three-component complex numbers (one real component, two complex components) but found there was no way he could multiply them in any meaningful way. It was only when he hit upon *quaternion* numbers, which took the form of having one real component and three 'imaginary' components,

$$q = w + xi + yj + zk,$$

that he hit the jackpot. In a quaternion number, w, x, y, z, are real numbers, and i, j, k, are imaginary numbers, each having the property that

$$i^2 = j^2 = k^2 = ijk = -1.$$

So instead of having just one imaginary number satisfying $i^2 = -1$, as is the case with complex numbers, we now have three such imaginary numbers. In order to make multiplication possible in this four-dimensional setting, we not only need to know what the product of the imaginary numbers is when multiplied by themselves, but also when multiplied by the other imaginaries. Hamilton's great stroke of brilliance are the products given above.

Such was Hamilton's excitement when he made this finding of how to multiply *quaternions,* on October 16, 1843, that he carved it into Broom Bridge (also Broome, Brougham) in Dublin (Fig. 3.5).

As we saw for complex numbers, the imaginary number i is associated with the ordered pair $(0, 1)$ in the two-dimensional Euclidean plane.

For the quaternions, the imaginary numbers i, j, k can also be associated with the standard unit vectors **i**, **j**, **k** in three-dimensional space, where $\mathbf{i} = (1,0,0)$, $\mathbf{j} = (0,1,0)$, and $\mathbf{k} = (0,0,1)$. These vectors have length one and point in the x-, y-, and z-directions respectively.

More generally, a quaternion $q = w + xi + yj + zk$, with the real term $w = 0$, is still considered a quaternion (called a *pure quaternion*) and is just the vector in three-dimensional space:

$$\mathbf{q} = (x, y, z) = x\mathbf{i} + y\mathbf{j} + z\mathbf{k}.$$

The class of all quaternions, pure and impure, is denoted by the symbol $\mathbb{H}$, for Hamilton.

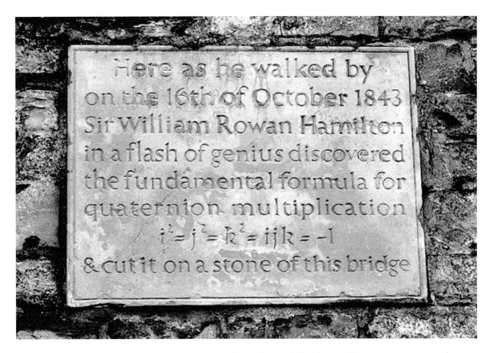

Figure 3.5: Affixed to the Broom Bridge in Dublin is a plaque commemorating Hamilton's act of vandalism as a result of his momentous discovery of how to multiply quaternions. The plaque reads: "Here as he walked by/ on the 16th of October 1843/ Sir William Rowan Hamilton/ in a flash of genius discovered/ the fundamental formula for/ quaternion multiplication/$i^2 = j^2 = k^2 = ijk = -1$ /& cut it on a stone of this bridge." Who says mathematicians are boring? (Image in public domain.)

In order to do addition or subtraction with quaternions, we simply add or subtract their respective components, as was done with complex numbers. It is the same procedure and the result is again a quaternion.

For multiplication of two quaternions, we need to have a complete list of the product of all the imaginary numbers, as in the following multiplication table (Fig. 3.6).

Note that to multiply two complex numbers, say

$$(a + bi) \times (c + di),$$

we could just employ the distributive law for multiplication, bearing in mind that $i^2 = -1$. Likewise for the multiplication of two quaternions,

$$(w + xi + yj + zk) \times (a + bi + cj + dk),$$

we also employ the distributive law, coupled with the values for the multiplication of imaginary numbers given in Fig. 3.6.

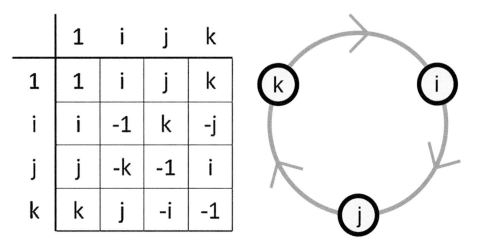

Figure 3.6: (L) Chart for the (noncommutative) multiplication of quaternions, with the number in the left column to be taken first; for instance, $ik = -j$, but $ki = j$. (R) More easily remembered is the cyclic nature of the multiplication, which is positive in the direction of the arrows and negative in the direction against the arrows. (Images courtesy of Wiki Commons.)

The result of the multiplication will be another quaternion. Some examples of quaternion multiplication are carried out in Appendices VII and VIII. The calculations can be somewhat tedious and messy as there are a lot of bits to multiply together and combine, but it can all be easily implemented on a computer.

The quaternion numbers were also known to the great German mathematician Carl Gauss (of course), but remained unpublished until after the work of Hamilton. One peculiar feature of quaternions, unlike the families of real or complex numbers, is that they do not in general *commute*, that is: $q_1 q_2 \neq q_2 q_1$.

For example, taking $q_1 = i$ and $q_2 = j$, then $q_1 q_2 = k$ but $q_2 q_1 = -k$, according to the multiplication table in Fig. 3.6. In the poetic words of mathematical physicist John Baez, "The quaternions, being noncommutative, are the eccentric cousin who is shunned at important family gatherings."

In spite of being the outcasts in the family, quaternions are especially useful when dealing with the rotation of an object in space, such as an airplane, rocket, or robotic arm, using computer graphics (Fig. 3.7). The reason is because a quaternion can be used to encode information about the *axis of rotation* as well as the *angle* θ to be rotated about this axis. This then leads to a mechanism for rotating an ordinary vector in three-dimensional space with a single operation involving quaternions. See Appendix VII for a worked example.

Quaternions also feature in other branches of Science, including Special Relativity and Quantum Mechanics. If only Hamilton were still alive to appreciate all the real-world applications his creation has led to. There would be no telling what he might carve next.

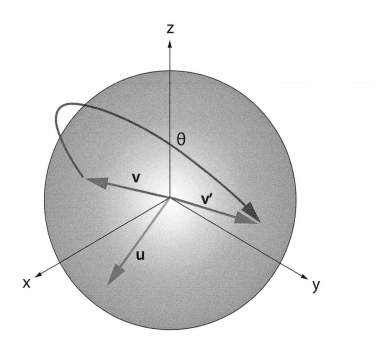

Figure 3.7: This diagram indicates a vector **v** (green) being rotated about an axis determined by a unit vector **u** (blue), through a given angle θ, to the new position **v'**, all of which can be achieved by a suitable quaternion operation. (Illustration courtesy of Katy Metcalf.)

One last remark before we leave the quaternions. Previously, we found that in the space of complex numbers, $\mathbb{C}$, there were two values that satisfied the equation $x^2 = -1$, namely, i and $-i$. Interestingly, in the space of quaternions, $\mathbb{H}$, it turns out that there are *infinitely* many such values, namely all the points in three-dimensional space that lie on a sphere of radius $=1$ (see Appendix VIII).

BUT WAIT, THERE IS MORE – OCTONIANS

Since quaternions have proved extremely fruitful, why stop with polynomial-like numbers having only four components? Actually, there is no need to, and there is one further number system we wish to discuss in which one can add, subtract, multiply, and divide. That system is called the *octonions,* $\mathbb{O}$[13].

---

[13] Actually, there are only four number systems where we can add, subtract, multiply, divide, and compute length in a suitable manner: $\mathbb{R}$, $\mathbb{C}$, $\mathbb{H}$, and $\mathbb{O}$. They are known as *normed division algebras* and have dimensions 1, 2, 4, and 8 respectively.

These were discovered by a friend of Hamilton's, John T. Graves, who was initially skeptical of what Hamilton had achieved and wrote to the former, "I have not yet any clear views as to the extent to which we are at liberty arbitrarily to create imaginaries, and to endow them with supernatural properties." But then Graves went on to create even more imaginaries — seven of them, in fact. However, his hitherto unpublished discovery was independently rediscovered by Arthur Cayley, who published about these numbers over a year later in 1845. They subsequently became known as 'Cayley numbers', but are now more commonly known as *octonions*[14].

Combined with a real component like the quaternions, the octonions also have a polynomial-like form:

$$x = x_0 + x_1 e_1 + x_2 e_2 + x_3 e_3 + x_4 e_4 + x_5 e_5 + x_6 e_6 + x_7 e_7,$$

where all the coefficients x_0, x_1, ..., x_7, are real numbers and the 'imaginary' numbers e_1, e_2, ..., e_7 all satisfy $e_i^2 = -1$, as well as the rules of multiplication given in Fig. 3.8.

$\times$	1	e_1	e_2	e_3	e_4	e_5	e_6	e_7
e_1		-1	e_3	$-e_2$	e_5	$-e_4$	$-e_7$	e_6
e_2		$-e_3$	-1	e_1	e_6	e_7	$-e_4$	$-e_5$
e_3		e_2	$-e_1$	-1	e_7	$-e_6$	e_5	$-e_4$
e_4		$-e_5$	$-e_6$	$-e_7$	-1	e_1	e_2	e_3
e_5		e_4	$-e_7$	e_6	$-e_1$	-1	$-e_3$	e_2
e_6		e_7	e_4	$-e_5$	$-e_2$	e_3	-1	$-e_1$
e_7		$-e_6$	e_5	e_4	$-e_3$	$-e_2$	e_1	-1

Figure 3.8: Multiplication table for the octonian units e_1, e_2,..., e_7, where the first number is to be taken in the left-hand column. Note that $e_3(e_4 e_5) = e_2 \neq (e_3 e_4) e_5 = -e_2$, so that octonian multiplication is not associative. (Illustration courtesy of Katy Metcalf.)

---

[14] Graves called his new eight-component numbers 'octaves'.

As we have seen, the quaternions are in general not commutative, but they are *associative*; that is,

$$q_1(q_2\,q_3) = (q_1\,q_2)q_3.$$

However, the octonions are neither commutative nor associative. This fact has put them out of favor with physicists. If quaternions are shunned at family gatherings, as noted above by John Baez, then octonians come in for even harsher treatment: "The octonions are the crazy old uncle nobody lets out of the attic: they are *nonassociative*."

Nevertheless, there is possibly a place for them in the physical world after all. By combining together the real numbers, $\mathbb{R}$, the complex numbers, $\mathbb{C}$, the quaternions, $\mathbb{H}$, and the octonians, $\mathbb{O}$, one arrives at a 64-dimensional space. Some mathematical physicists like Cohl Furey think perhaps this is a suitable mathematical structure that underlies the strong, weak, and electromagnetic forces of the Standard Model of the Physics of elementary particles.

THE WORLD'S HARDEST PROBLEM – THE RIEMANN HYPOTHESIS

If I were to awaken after having slept for a thousand years, my first question would be: Has the Riemann hypothesis been proven?...
Mathematician David Hilbert

Just stretching our minds, a little, we can discuss what is arguably the world's hardest and most famous mathematical problem. Although we can never be sure of such a claim, it has stumped the greatest mathematical minds since it was first propounded in 1859 by Bernhard Riemann, whom we encountered earlier. We are already familiar with the basics now, so not much is new here, but it is further discussed in Appendix IX.

Let us just begin by stating that, just as for real variables, we can consider functions with complex variables. For example, we discussed the function,

$$y = \frac{1}{x^2},$$

in Chapter 2. Plug in any value for $x \neq 0$, and the function outputs a corresponding value for y. Considering all values of $x \neq 0$ gives us the graph of the function (Fig. 2.1, Chapter 2). The function displayed a singularity at $x = 0$, as the value of y became infinite as x approached the origin.

Similarly, we can consider functions involving complex variables, such as

$$w = \frac{1}{z^2},$$

where the variable z takes on all values in the complex plane, except where w is *not defined* at $z = 0$, which again is a singularity for this complex function. For all the complex values of z, there will be a specific complex value w determined by the equality. Functions are like machines: give it some input and it will generate suitable output calculated from the expression defining the function.

The main difference occurs when considering the graph of complex functions. The values of z lie in the complex plane but the complex values of w also lie in a plane. One way to achieve a three-dimensional graph is simply to consider the modulus of w, $|w|$, which is a *real* number and its values can be taken as those of a third height axis above the plane.

So, for every point z in the plane, there is a corresponding point w given by the function, but instead we take the value of $|w|$ to give the third coordinate in the three-dimensional representation. It is a worthwhile fudge that gives a feeling for how the function is behaving, in the sense of the output getting closer or further from the origin as z roams around the complex plane. (See Fig. 3.9).

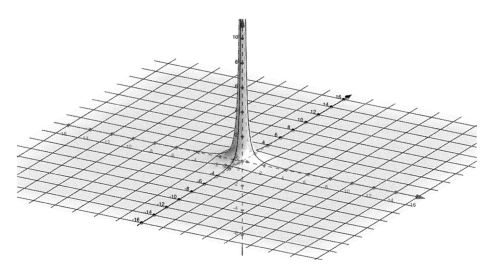

Figure 3.9: A three-dimensional graph associated with the function $w = \dfrac{1}{z^2}$, for values of $z \neq 0$ in the complex plane, where the height oordinate is given by the real number $|w|$. Red is the real-axis, green the imaginary axis. Compare this with graph of the function $y = 1/x^2$ in Fig. 2.1 in Chapter 2. Both functions have a singularity at the origin, becoming infinitely large there. (Image courtesy of GeoGebra.org)

Before we tackle one of the most pre-eminent of all functions, let us go back to the infinite series in eq. (5) for a moment:

$$\frac{1}{1^2} + \frac{1}{2^2} + \frac{1}{3^2} + \frac{1}{4^2} + \frac{1}{5^2} + \cdots = \frac{\pi^2}{6}.$$

This series was spawned from the harmonic series eq. (4), but here the terms in the denominator are all powers of 2. Taking powers of the counting numbers (like 2) in the denominator makes the series converge, unlike the harmonic series which diverges to infinity.

Instead of raising the counting numbers in the denominator to the power 2, if we take the power 4, we have another beautiful infinite sum involving the value π, but like the preceding sum there is no apparent connection with circles:

$$\frac{1}{1^4} + \frac{1}{2^4} + \frac{1}{3^4} + \frac{1}{4^4} + \frac{1}{5^4} + \cdots = \frac{\pi^4}{90}.$$

Indeed, taking any power greater than 1 of the counting numbers in the denominator will make the series converge, though not necessarily to such beautiful expressions as those above involving π. So, we can say that whenever the power $p > 1$, the series (called a *p-series* for obvious reasons):

$$\frac{1}{1^p} + \frac{1}{2^p} + \frac{1}{3^p} + \frac{1}{4^p} + \frac{1}{5^p} + \cdots,$$

will converge to some finite value. Of course, whatever value the series converges to depends on the value of p. When $p = 2$, the series converges to the value $\frac{\pi^2}{6}$. When $p = 4$, the series converges to the value $\frac{\pi^4}{90}$. The reader might wish to try summing the series for $p = 6$[15].

Now let us make a giant leap for mankind with the great mathematician Bernhard Riemann. Replace the *real* number power p in the p-series expansion with a *complex* number power, and call the power s[16]. Unsurprisingly, the series does not look all that different:

$$\frac{1}{1^s} + \frac{1}{2^s} + \frac{1}{3^s} + \frac{1}{4^s} + \frac{1}{5^s} + \cdots$$

---

[15] For $p = 6$, the p-series sums to $\frac{\pi^6}{945}$.

[16] We have been using the letter z for complex numbers, but the letter s is traditional in this context, so let us go with the flow.

The only difference is that the real number power p has been replaced by the complex number $s = x + iy$. All in all, this series looks just like the preceding one involving p, and is not too scary.

Is that all there is to it, just replacing a real exponent p by complex exponent s? Well, almost but not quite. First of all, we must again know where this series converges. As mentioned above, the ordinary p-series converges for any value of p greater than 1. For this new series to converge, the complex numbers $s = x + iy$ must all have their real part subject to the constraint that $x > 1$, which is analogous to $p > 1$[17].

Furthermore, note also that the preceding infinite sum with complex powers s behaves like a function. That is to say, input a value $s = x + iy$ (subject to $x > 1$) and the sum will give an explicit output value. We could even put in the value $s = 2$ and we already know that the output value would be $\dfrac{\pi^2}{6}$, or for $s = 4$, the output value would again be as before, $\dfrac{\pi^4}{90}$. In general, since $s = x + iy$ is a complex number, the output will also be a complex number.

Our infinite sum involving the power s is actually a function, whose value depends solely on the input value of s, so let us write this function with its customary name:

$$\zeta\left(s\right) = \frac{1}{1^s} + \frac{1}{2^s} + \frac{1}{3^s} + \frac{1}{4^s} + \frac{1}{5^s} + \cdots \tag{13}$$

Here, the letter ζ is the Greek letter *zeta* 'z'. Thus, for any s in the allowable domain where the sum converges, the infinite sum of eq. (13) will crank out a value which we call $\zeta(s)$. Every time.

We have nearly arrived at the promised land. Of course, the point $s = 1$ is a dangerous one that makes the function blow up to infinity as eq. (13) becomes the harmonic series. So, this one point is a singularity and we simply avoid it.

But what about all the other values in the complex plane $s = x + iy$? So far, we only have our function $\zeta(s)$ defined on the right of the vertical line $x = 1$, when the real part x of s is greater than 1. That is not such a big achievement as it is just the p-series with the real number p replaced by the complex number s.

So what?

What we really seek next, and what Riemann achieved, is to extend the scope of the function $\zeta(s)$ in a suitable natural manner to all the numbers $s = x + iy$ in the *entire* complex plane (while still avoiding $s = 1$). These values are derived from the values already defined by eq. (13) in a very specific manner and this is done in Appendix IX.

---

[17] For a complex number $s = x + iy$, the constraint $x > 1$, means that s is free to roam in that region of the complex plane whenever the real part of s satisfies $x > 1$. This is a half-plane to the right of the vertical line $x = 1$.

This *extended* function to the whole complex plane is *still referred to as* $\zeta(s)$, and is known as the *Riemann zeta function*. We have now reached the promised land. From its description in Appendix IX, it turns out that this has a completely new formulation to the one we started with in eq. (13). However, a most important feature of the Riemann zeta function $\zeta(s)$ is that it agrees exactly with the values of $\zeta(s)$ as determined initially by eq. (13) whenever $x > 1$.

So, for example,

$$\zeta(2) = \frac{1}{1^2} + \frac{1}{2^2} + \frac{1}{3^2} + \frac{1}{4^2} + \frac{1}{5^2} + \cdots = \frac{\pi^2}{6},$$

and $\zeta(4) = \dfrac{\pi^4}{90}$, for the extended $\zeta(s)$, just as before. Moreover, the Riemann zeta function is very well-behaved in the entire complex plane, except at that one unruly point $s = 1$, where the function has a singularity.

At this stage, we are particularly interested in those values of s in the complex plane where the Riemann zeta function outputs the value zero; that is, $\zeta(s) = 0$. Any point at which a function outputs the value zero is known as a *zero* of that function (surprise). For example, the point $x = 1$ is a zero of the function $y = x - 1$.

Much time is spent in high school finding the *roots* of polynomials[18] and these are just the points (the zeros) where the polynomial function takes the value zero, or in other words, where the graph of the polynomial crashes into the x-axis.

According to the definition given in Appendix IX, which we need not go into here, the new extended Riemann zeta function happens to output the value zero at all the even negative integers; that is,

$$\zeta(-2) = \zeta(-4) = \zeta(-6) = \ldots = 0.$$

But because these points where $\zeta(s) = 0$ simply arise from the definition of $\zeta(s)$, they are known as *trivial zeros*. We are not in the least bit interested in these.

On the other hand, it has been found from calculations that the Riemann zeta function also outputs the value zero at infinitely many other points in the complex plane. Riemann himself calculated a few of these zeros, noticing that they all have real part $x = 1/2$. How curious. See Fig. 3.10 for the location of the first few of these 'nontrivial' zeros.

---

[18] Recall the dreaded quadratic formula from high school:

$$x = \frac{-b \pm \sqrt{b^2 - 4ac}}{2a},$$

which gives the roots of the polynomial $y = ax^2 + bx + c$.

Since we will be dealing with infinitely many of these zeros, let us use some convenient notation just to keep track of them. The known (nontrivial) zeros are commonly written as:

$$s_n = \frac{1}{2} + i\,t_n,$$

where the real part is 1/2, and the imaginary part is going to be some decimal number that is getting increasingly larger the more zeros we calculate.

Because of the symmetry in the way the Riemann zeta function is defined, each of these zeros has a counterpart in the complex plane below the x-axis, at

$$s_n = \frac{1}{2} - i\,t_n.$$

But since we know they are always going to be there we can safely ignore them, as they are all coupled with the ones above the x-axis. That is where the action is.

Riemann's sublime conjecture, the *Riemann Hypothesis* (RH), which has a price on its head of US$1 million, is that *all* the (nontrivial) zeros s_n of the zeta function lie on the *critical line* x = 1/2 (Fig. 3.10).

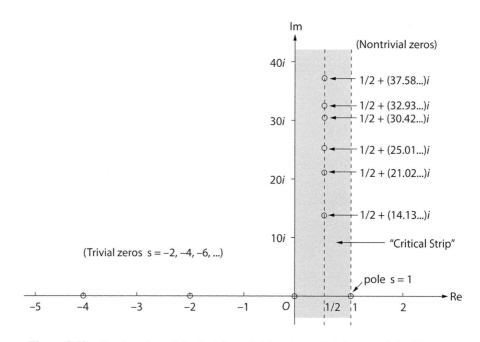

Figure 3.10: The location of the first few trivial and nontrivial zeros of the Riemann zeta function, the latter of which all have real part x = 1/2. The singularity occuring at s = 1 is known as a *pole*. (Illustration courtesy of Katy Metcalf.)

Indeed, the conjecture is arguably the most important unresolved question in all of Mathematics, certainly one of the most famous, and undoubtedly one of the most fiendishly difficult.

The hypothesis is likely the most obscure thing that the reader has ever encountered in their life. While seemingly not terribly exciting, it is actually extremely important as it has deep connections with the distribution of prime numbers, and it even has connections with Quantum Mechanics. There is a very beautiful direct connection between the Riemann zeta function and all the prime numbers that is given by the famous Euler product formula, also discussed in Appendix IX.

Furthermore, Riemann had found an explicit formula for determining the number of prime numbers up to any given number, as well as measuring their distribution, but the formula involved taking a sum over the nontrivial zeros of his zeta function. The proviso for the formula to work was that all the nontrivial zeros s_n had to lie on the critical line $x = 1/2$.

In view of its significance to Mathematics, many great mathematicians have endeavored to make some headway toward a proof ever since it was first enunciated by Riemann. The distinguished English mathematician, Godfrey Harold (G.H.) Hardy (1877–1947) proved in 1914 that there are *infinitely many* zeros on the critical line $x = 1/2$. Taking a different approach, Andrew Odlyzko has used a computer to find the first ten trillion zeros of the zeta function, and sure enough, they all lie on the critical line[19].

Odlyzko has also computed the first 100 zeros to an accuracy over 1,000 decimal places. On the other hand, all one needs to do to *disprove* the Riemann Hypothesis (RH) is to find a single zero of the zeta function that does *not* lie on the critical line. Many books have been written about the Riemann zeta function and the hypothesis concerning the location of the zeros (see John Derbyshire's excellent book in the Bibliography). In another vein, it was shown in 1974 that at least one-third of *all* the nontrivial zeros lie on the critical line, and 15 years later that number was marginally increased to two-fifths.

Strangely enough, another approach to proving RH is via Quantum Mechanics. There is evidence that indicates a correspondence between the distribution of the zeros of the Riemann zeta function along the critical line and the distribution of certain real numbers (called *eigenvalues*) associated with a particular class of random matrices (called *Hermitian*). Since the latter are directly related to quantum mechanical states, two mathematical physicists, Michael Berry and Jonathan Keating, have conjectured[20] that the nontrivial zeros s_n of the zeta function exactly correspond to the energy levels E_n of some quantum system, so that

$$s_n = \tfrac{1}{2} + i\,E_n.$$

---

[19] http://www.dtc.umn.edu/~odlyzko/zeta_tables/index.html

[20] The Riemann Zeros and Eigenvalue Asymptotics, *SIAM Review*, Vol. 41 No. 2, pp. 236–266, 1999.

The distinctive feature of their proposal is that the (still unknown) quantum system possesses a classical counterpart whose orbits are chaotic[21], thereby linking three areas previously regarded as separate: Arithmetic, Quantum Physics, and Chaos Theory.

  If such a quantum system could be found, then RH would be proved. This is so because the energy levels, E_n, are all physically real numbers and so all zeros, s_n, must necessarily lie on the critical line[22].

  On the other hand, many have found other mathematical results that are completely equivalent to the Riemann Hypothesis, so if any one of these is proven, then so is RH.

  One interesting value that is outputted by the Riemann zeta function is when $s = -1$, namely,

$$\zeta(-1) = -1/12.$$

We have seen this value of $-1/12$ before, for if we surreptitiously sneak the value $s = -1$ into the original series formulation of the zeta function (eq. 13), then all the numbers formerly in the denominator pop up into the numerator and we get the result:

$$1 + 2 + 3 + \ldots = \zeta(-1) = -1/12.$$

This is the equality depicted in the painting of Fig. 2.6 in Chapter 2, which employs a somewhat different formulation than the one in eq. (35) of Appendix IX for the Riemann zeta function. Of course, we have overstepped the mark here in putting in the value $s = -1$ into the original representation for $\zeta(s)$, as that expression was *only* valid in the half-plane when $x > 1$ ($s = x + iy$), and thus is not valid in the left half of the complex plane. Evidently, however, in the world of String Theory, our little illegitimate step of using eq. (13) for the value $s = -1$ seems to make some kind of sense.

  We have just taken a journey into one of the most far-reaching corners of the Mathematical Universe and hopefully the reader has survived with all faculties intact. The journey will not be so bumpy from here on out.

---

[21] Chaotic behavior and orbits are explained in the section on Dynamical Systems in Chapter 6.

[22] This is a refinement of the so-called Hilbert-Pólya conjecture, essentially due to Pólya and possibly due to Hilbert, which says that the zeros of the zeta function are the eigenvalues of a certain type of 'operator'. More in Chapter 7 on operators.

4

Random Universe

We are the cosmos made conscious and life is the means by which the universe understands itself...
Physicist Brian Cox

So one reason why scientists do mathematical theories is because they surprise us, they become smarter than us, and eventually we become the students of the theory.
Cognitive Scientist Donald Hoffman

In this chapter, we are going to examine some random events in the real world. In a sense, we will try to tame randomness with Mathematics and see what that tells us about our Universe.

GOING STEADY

In Chapter 1 (the section entitled 'Off to Monte Carlo') we saw that randomly dropping grains of sand on a square with an inscribed circle led to a determination of the value for the number π. In fact, randomness plays a fundamental role in solving many problems in the physical world.

One interesting application of randomness, which gives an answer to a concrete real-world problem, would be determining the heat distribution in a closed region when you know the temperature on the boundary of the region. For simplicity, let us consider this problem in two-dimensions and make the region circular. Let us further suppose that the boundary of the circle is an arbitrarily thin iron ring.

© Springer Nature Switzerland AG 2020
J. L. Schiff, *The Mathematical Universe*, Springer Praxis Books,
https://doi.org/10.1007/978-3-030-50649-0_4

Now, let us heat one portion of the ring and let the whole ring come to an equilibrium (*steady-state*[1]) temperature, so that one portion is hot, and further away the ring is cooler. We are interested in the temperature at points inside the ring.

This is what mathematicians refer to as a *Dirichlet Problem*, whereby a continuous temperature distribution (some function) is put on the boundary of a region and the question is asked: 'What is the resulting equilibrium temperature distribution in the region inside the boundary'? The solution to this problem has an answer that is known as a *harmonic function*. Without going into a discussion of these functions, they have one extremely useful natural property[2].

This notion of the Dirichlet Problem is somewhat analogous to the Holographic Principle discussed in Chapter 1. All the information concerning, say, the temperature at each point inside a region, is encoded in the temperature values on the boundary.

What would you expect the temperature to be at the *center* of the circular region, just using some common sense? Let us see.

Suppose the boundary ring is hot along one section, then gets much cooler as you go around to the far side of the ring, before becoming warmer and warmer on your return to the starting point. The point at the center of the ring is equidistant from every point on the ring boundary (Fig. 4.1). Therefore, each boundary temperature value will have the same influence on the temperature of the central point as any other, and thus the central point must accommodate *all* of the varying temperatures on the boundary.

This is just what the harmonic function solution to the Dirichlet Problem takes into account, and perhaps the reader has already guessed the answer. The temperature at the center is exactly the *average* of all the temperatures on the boundary, or what mathematicians call the *mean-value property*. The same thing would hold in three-dimensions if we applied a continuous temperature distribution on a spherical surface. The steady-state temperature at the center of the sphere would be the average of all the temperatures on the boundary surface. Mother Nature is taking exactly what seems to us to be the most natural option.

Mathematically, of course, we have an infinite number of points on our ring boundary, and a bit of real Mathematics needs to be performed to sum them up to find their average, but this can be done. Moreover, we can even allow for a finite number of jumps in the temperature, known as *discontinuities*. For example, suppose that the temperature is 100° on one half of the ring and 0° on the other half. Thus, the temperature at the center of the ring would be 50°.

---

[1] This means that the temperature does not vary over time.

[2] Actually, harmonic functions have many useful properties, but we only mention one of them here.

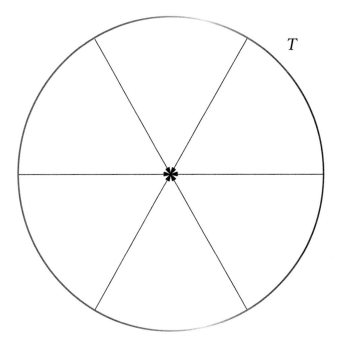

Figure 4.1: The temperature at every point on the boundary has an equal influence on the temperature at the center. (Illustration courtesy of Katy Metcalf.)

But what does the temperature distribution at all points inside the circular region actually look like if we did have a 100° and 0° temperature distribution on each half of our circular boundary, for example? One method would be to find the harmonic function solution to the problem mathematically and then we could evaluate this function to find the temperature at any particular point.

But there is another approach, one that entails taking a drunken walk from a point inside the region. This is called a *random walk*. This will be a journey taking very small steps, starting from the point we wish to determine the temperature at, say P, so that the very next step is always taken in a completely random direction. Such an approach is perfect for a computer simulation, as in Fig. 4.2.

The random walk starts at our point P in question and carries on until eventually it hits the boundary[3].

According to a rather remarkable mathematical result discovered by the Japanese-American mathematician Shizuo Kakutani in 1944, the probability (times 100) of the random walk starting at point P and *first encountering the boundary* where the temperature is 100° is equal to the value of the harmonic function at the point P, and thus gives the temperature at P. (Recall that we have already come across Kakutani in Chapter 1 with regard to his fixed-point theorem, used by John Nash to prove the latter's famous equilibrium theorem.)

---

[3] It is known that the random walk will eventually hit the boundary at some stage.

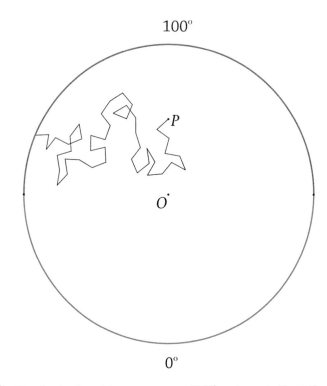

Figure 4.2: Our circular ring with a temperature of 100° on the top half and 0° on the bottom half. We would like to find the temperature at the point P. It can be found by starting a random walk at P and letting it continue until it hits the boundary. Repeating this operation multiple times, one can compute the probability of hitting the 100° side first and multiplying that value times 100 will determine the temperature at P. (Illustration courtesy of Katy Metcalf.)

This makes intuitive sense, because if the point P is near the segment where the temperature was 100°, then the probability of hitting that segment first would be much higher than hitting the 0° segment further away. This higher probability is reflected in the higher temperature at P.

Similarly, if our starting point P is near the 0° region, the probability of hitting the 100° segment first would be low and this would be reflected in the low temperature there. It all makes perfect sense.

But how would we work out this probability in practice. Quite simply, we start at our chosen point P and let the computer generate a random walk from that point until it hits the boundary circle. As soon as the boundary is hit, the computer generates another random walk from point P, and so on. With each random walk, we keep track of whether it hits the 100° segment or the segment with 0°. We can do this simulation 1,000 times or 10,000 times, the computer does not mind.

Then, the number of times our random walk hit the 100° segment, divided by the total number of random walks taken, is the probability of hitting the 100° segment first. Multiplying this probability times 100 gives us the temperature at P.

Who would have thought that a drunken walk could lead to something so remarkable as determining the temperature at a given point in an enclosure?

As an example, let us consider a circular region of radius 1 unit, and keep the top half of the boundary at $100°$ and the bottom half at $0°$.

Suppose we want to find the temperature at a point P half way up from the center to the top boundary, as in Fig. 4.2[4], by taking repeated random walks from P as described above.

The results from taking 1,000 and 10,000 simulations respectively were: $79.2°$ and $79.3°$. This compares very favorably with the mathematically computed value of $79.5°$. See Appendix X for the computer code so you can run the simulation yourself.

BROWNIAN MOTION

Such random walk behavior was also noted by the eminent Scottish botanist Robert Brown in 1827, when observing small pollen grains under a microscope while they were suspended in a water droplet. Brown observed the random movement of microscopic particles within the pollen grains and concluded that the motion was not due to any external forces but "belonged to the particle itself." This movement is now termed *Brownian Motion*, although it had already been discovered by Dutch physician and scientist Jan Ingenhousz in 1785, who described the erratic motion of coal dust on the surface of alcohol. Indeed, Brown asserted that "any solid mineral would reveal the phenomenon subject to its being reduced to a sufficiently fine powdery form[5]."

The scientific explanation was only given in 1905 by Albert Einstein, who showed that the motion of microscopic particles in a liquid was the result of the random movement of the water molecules that collided with the particles. This process is related to that of *diffusion*, as, for example, putting a drop of ink into a bowl of water. The ink becomes diffused by the Brownian Motion from the water molecules. Indeed, that is conceptually how we found the temperature at the given point in the preceding example, by simulating the random Brownian Motion of an air molecule.

LIFE IS A GAMBLE

Random walk models also feature in many other scientific fields, such as Physics, Biology, and Economics. From a personal economic standpoint, you might wonder why, when you play a game like blackjack at a casino which gives you a near

---

[4] This is just the point (0, 0.5) in the plane.

[5] Brian J. Ford, Brownian motion in *Clarkia* pollen: A reprise of the first observations, *The Microscope*, 40 (4): pp. 235–241, 1992.

even chance of winning, you still tend to walk away empty handed[6]. Unless of course you quit while you were ahead.

The reason is because of another walk, actually another a random walk in this case. Whatever your stake is — say $100 — and regardless of whether you bet the same or varying amounts on each hand, then a graph of your total revenue can still be modelled by a random walk, as in Fig. 4.3. Mathematics says that your revenue will return to $0 in the long term. In fact, if the game went on forever, it would return to $0 infinitely often, but this is only the theory and the first time your holdings return to $0 is sufficient to put you out of the game.

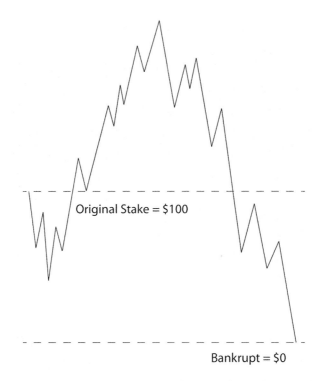

Figure 4.3: A graph depicting the random walk nature of a gambler's total wealth, starting with $100, at any game with equal odds of winning at a casino. The crunch comes because the gambler has a finite amount of wealth and the casino has a virtually infinite amount. So eventually your wealth will hit $0 and it is game over for you. In addition, most casino games do not quite give you equal odds, so this graph presents a generous picture of your long-term earnings. (Illustration courtesy of Katy Metcalf.)

---

[6]The odds when playing blackjack are about 0.5% in the casino's favor over the long term, but we can essentially take the chances of the casino winning as equal to that of the punter.

The basic problem is that the casino has a virtually infinite stake to gamble with, while yours is always finite. This is one version (among others) of what is called 'gambler's ruin' and the title is well deserved. So, when you walk away from the blackjack table having lost your entire stake, there is no reason to think that the dealer was any smarter than you or that you have been conned. You were undone by the mathematical inevitability of a random walk.

The moral of this story is to leave the table while you are ahead, but of course this is easier said than done, and once you are on a 'winning streak' (again, random inevitability) you want to continue. You do so at your peril.

EXPONENTIAL DECAY

In contrast to losing money by gambling, we discussed in Chapter 2 that your money could grow exponentially via the formula for continuously compounded interest given by a bank.

It is not only your money that can increase in this fashion; populations can do so as well, including human, animal, insect, and bacterial. The exponential function, e^x, is involved in the study of all of these. In addition, cables that are supported at each end and sag under their own weight will form the shape that is known as a *catenary,* which is also defined by exponential functions. The exponential e is as much a part of our Universe as the ubiquitous π, perhaps even more so.

So instead of exponential growth, let us turn to the opposite phenomenon of the way many things decay with time, known as *exponential decay.*

For example, if you have a quantity of a substance, such that every eight days half of it is gone or replaced by something else, then the situation could be graphed as in Fig. 4.4. This actually turns out to be the situation with the radioactive substance iodine-131, which is used in the treatment of thyroid cancer. It also occurs in nuclear reactions and was one of the health hazards resulting from the Chernobyl and Fukushima disasters.

Interestingly, if one watches an individual Iodine-131 atom, there is no way to predict when it will decay. The decay of the atom will be entirely random and could take place in an hour, six months or 100 years. Yet an aggregate of iodine-131 will be half decayed in eight days. This is rather remarkable if you think about it. Random at the individual atomic level, but quite regular when dealing with an entire mass.

Radioactive decay follows a fundamental rule, in that the rate of decay is proportional to the amount of the substance that is present at any given time. This leads to an equation involving the exponential function, which in turn allows us to use radioactive decay as a clock that works backwards in time.

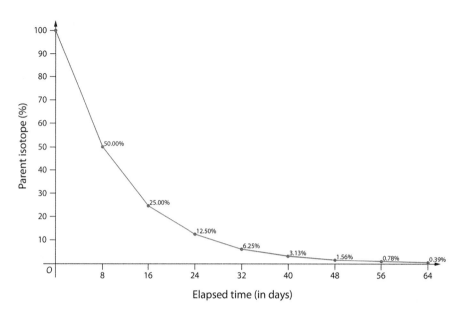

Figure 4.4: A depiction of the situation when half of a given substance decays into something else every eight days. The remaining percentage of the substance is indicated over time. In fact, this is the graph of the decay of radioactive iodine-131, whose shape is typical of exponential decay. (Illustration courtesy of Katy Metcalf.)

THE DATING GAME

There are various substances that are known to be radioactive, including radium, uranium, carbon-14, and many others. Because these substances decay into other substances at a fixed rate, like the ticking of a clock, they can be used for dating purposes. To make use of this radiometric clock, we need to know the rate of decay, which is conveniently described by the substance's *half-life*; that is, the time it takes for half of any given quantity of the substance to decay.

For example, a form of uranium (U-238) has a half-life of 4.468 billion years, decaying into a stable non-radioactive form of lead (Pb-206). Therefore, if you knew that a half of the original quantity of uranium in a rock sample had decayed to lead, then your rock sample would be 4.468 billion years old.

On a shorter time-scale is a form of carbon (C-14) that is produced in the Earth's upper atmosphere by cosmic rays. C-14 is also radioactive, with a half-life of 5,730 years, and decays into a stable form of nitrogen (N-14) that is not radioactive. Because living things absorb carbon-14, radio carbon dating of ancient artifacts up to about 40,000 years old has become an important branch of archeology and anthropology. For older objects, other methods are employed.

Let us be rather ambitious and date our Solar System. This formed out of a flattened swirling cloud of gas and dust (the *Solar Nebula*) and at the center was the proto-Sun. As the gas and dust in the cloud cooled, material began to solidify out of the cloud and gradually formed the planets. Out between Mars and Jupiter lies a rotating cloud of solidified masses known as the Asteroid Belt. Tidal forces created by Jupiter's enormous gravity were too great for any planets to form in this region and instead we are left with hundreds of thousands of planetesimals, or asteroids (Fig. 4.5).

These asteroids are of various sizes, with the largest being Ceres at 946 km and which has now been upgraded to the status of dwarf planet. Being very numerous, these asteroids at times collide with one another and, due to certain dynamical forces, collisional bits of debris can land on Earth. These are known as *meteorites*. Some asteroid debris has hit the Moon, as the many impact craters on its surface bear witness, while more than 120 impact craters have been found on Earth.

Figure 4.5: The asteroid Vesta, the fourth to be discovered, with a diameter of 530 km. Note all the craters on the surface that are the result of collisions with other asteroids. Some fragments of prior collisions with Vesta have landed on Earth and have been analyzed in laboratories. (Image courtesy of NASA/JPL.)

Unlike the rocks on Earth, most asteroid fragments have not experienced any weathering and have experienced little geologic processing. Some asteroids have had volcanic activity, but as they are small bodies they have completely cooled down, unlike the Earth which still has a molten core. The non-volcanic material from the surface of an asteroid is very pristine, essentially unchanged since it was formed billions of years ago. Therefore, if we have such material from an asteroid, we can analyze it in a laboratory for its radioactive constituents and find the age of the asteroid, which will be the age of Solar System.

To date the Solar System, nothing more than some high school algebra is required, along with the general formula for exponential decay:

$$N = N_0 e^{-kt}. \tag{14}$$

Here, N_0 represents the initial amount of the decaying quantity, represented by the 100% in the graph above (Fig. 4.4). The constant k is particular to the specific decay process – as every substance decays at a different rate – and is derived from the half-life of the substance[7]. The variable t is just the elapsed time we are considering, and the quantity N is then the remaining quantity at that particular time t. This is the standard equation that governs the behavior of many decay processes in the natural world and in the Universe at large.

Back now to dating the Solar System. We will sketch the procedure here, but provide all the details in Appendix XI.

Suppose we have a meteorite sample from the Asteroid Belt and we analyze it for the element rubidium, which has a convenient half-life. A radioactive form of rubidium (Rb-87) decays into a stable form of the element strontium (Sr-87), and the half-life of this decay is 49.5 billion years. Using the above eq. (14), now with respect to rubidium, it looks like this:

$$^{87}Rb = {}^{87}Rb_0 e^{-kt},$$

where $^{87}Rb_0$ denotes the initial amount of rubidium that was in our meteorite sample when the asteroid formed, and ^{87}Rb is the value of the amount that is currently present. The value k is a known decay constant for rubidium, and our task is to find the time t that has elapsed since the rubidium was formed in the asteroid. This will give the age of the Solar System.

We would ultimately like to solve the preceding equation for the value t. However, a missing ingredient is the initial amount of rubidium that was present, namely, $^{87}Rb_0$, which we do not actually know the value of since no one was around at the time to measure it. It would seem that we are at an impasse.

---

[7]If $T_{\frac{1}{2}}$ denotes the half-life of a radioactive substance, the decay constant k is given by:

$k = \left(1/T_{\frac{1}{2}}\right)\ln 2 = 0.693\left(1/T_{\frac{1}{2}}\right).$ This is derived directly from eq. (14).

Yet all is not lost, since we can measure the amount of the decay product strontium-87 that is in the meteorite, and with a bit of high school algebra, a clever trick, and nothing more than the equation of a straight line, the age of the Solar System reveals itself. It works out to 4.56 billion years (the details provided in Appendix XI are not difficult), and this is the age of the Earth, planets, asteroids, and Sun.

Just as in exponential decay, another simple model − this time of exponential growth − is that of a population of living organisms. A simple model is one in which the rate of population increase is proportional to the number of individuals, N, present in the population at a given time, t. The appropriate equation is just a variation in sign of the model for exponential decay,

$$N = N_0 e^{kt},$$

where k is a positive constant that depends on the growth rate of the population and N_0 is the value of the initial population. In studies of real populations, the value of N_0 is easy to obtain, making this an easier calculation than working out the age of the Solar System.

All this can be derived using the remarkable number e.

EMPOWERING LAWS

Growth and decay behavior that is superficially similar looking to the preceding, involving the exponential e, is described by what is called a *power law*; that is, where one quantity varies according to the power of another. There are many models of natural phenomena that depend not on the exponential e, but on some positive or negative parameter a, but in this instance as a power.

Our power law formula is,

$$y = kx^a,$$

for $x > 0$, with a, and k, constants.

The common parabola $y = x^2$ is a familiar example, but let us turn our attention to when the constant a is negative.

One example is avalanches. Small ones are fairly common, but larger and larger ones become ever more infrequent. Another, rather sobering example of a power law is the size of the asteroids that strike the Earth. Small asteroid fragments hit the Earth every day[8], but fortunately larger objects from space impact the Earth less frequently. If we graph the number of impacts per year versus the size of the object, then the graph will look something like Fig. 4.6.

Suppose we have a graph of data as in Fig. 4.6, and suspect it is following a power law. How do we find the values of k and a? There is one little mathematical

---

[8] Almost invariably, these originate in the Asteroid Belt between Mars and Jupiter and, through dynamical processes, find their way to Earth as mentioned in the text.

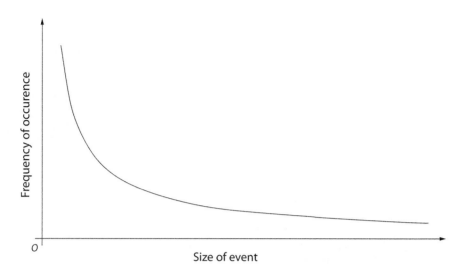

Figure 4.6: A typical power law graph of certain natural phenomena. If the frequency of occurrence is represented along the vertical axis and the size of the event along the horizontal axis, then small events occur frequently and large events occur rarely. (Illustration courtesy of Katy Metcalf.)

trick that is employed for doing this. In a true power law relation, of the form $y = kx^a$, taking the logarithm of both sides (often base 10, which also has the effect of keeping large numbers manageable), yields,

$$\log y = \log k + a \log x,$$

or, in a more familiar form,

$$\log y = a \log x + \log k.$$

Why have we taken the logarithm of both sides? The reason is that this equation now has the form of a straight line:

$$Y = MX + B,$$

that we all saw in high school. Here, $Y = \log y$, the slope is $M = a$, the independent variable is $X = \log x$, and $B = \log k$.

So, any power law equation will be turned into a straight-line equation, with a *log-log* graph representation; that is, $Y = \log y$ plotted against $X = \log x$. The all-important value of the power a is given by the line's slope M, which is easily measured on a graph, and $k = 10^B$ derives from the y-intercept, B. Such *log-log* graphs are common in many branches of Science and also feature in the measurement of the size of supermassive black holes at a galaxy's core, as will be shown in Chapter 6.

Going back to asteroid impacts, their size can be calculated from the bright flash of the detonation that is picked up by orbiting satellites as the asteroid hits the Earth's upper atmosphere, as well as from infrasound[9] stations that are part of the nuclear test ban verification system. Since very small asteroid fragments impact every day and are nothing to worry about, let us just consider the bigger ones that are of more concern.

These start with garage-size asteroid fragments that turn up at the top of our atmosphere on average once a year, possessing an energy of around 4 kilotons worth of TNT. The majority of small asteroid fragments (technically known as *meteoroids*) burn up harmlessly in the atmosphere without causing any damage. But a few that do get through the atmosphere will either end up in the oceans which comprise 70% of the Earth's surface, or end up on land as *meteorites*. Some will have even originated from asteroid Vesta.

Over a more than eight-year study of meteoroids impacting the upper atmosphere, it was found that the cumulative number, N, of impacts per year, in the $1 - 200$ meter range, followed the straight line log-log relation,

$$\log N = -2.70 \log D + 1.57,$$

where D is the diameter of the meteoroid[10]. Here we have evidence of a power law relationship between the frequency of impacts and the size of the impactor.

The largest asteroid event in recent times was the Chelyabinsk fireball that passed over the southern Ural region of Russia on February 15, 2013, possessing an energy of $400 - 500$ kilotons of TNT (see Appendix XII where this is calculated). The airburst caused numerous injuries, mainly from windows blown out by the shock wave, as well as damage to thousands of buildings. The Chelyabinsk meteoroid is the largest (at about $17 - 20$ meters across) to penetrate the atmosphere since the Tunguska event, also over Russia, in 1908 (Fig. 4.7). This latter event flattened some 2,000 square kilometers of forest in an airburst having an energy of about 3–6 megatons of TNT[11].

If we put a 20-meter diameter meteoroid into the preceding equation, then the value of $\log N$ is almost -2, which means that this was close to a one in a hundred-year event. Chelyabinsk (Fig. 4.8) was just such an event.

A Tunguska-size event is expected very roughly once every 1,000 years or so. At the far end, we have an event that did occur 66 million years ago, involving an

---

[9] Low frequency sound waves.

[10] P. Brown, *et al.*, The Flux of Near-Earth Objects Colliding with the Earth, *Nature,* 420, 294–296, 2002.

[11] For comparison, the atomic blast of Hiroshima was about 15 kilotons of TNT.

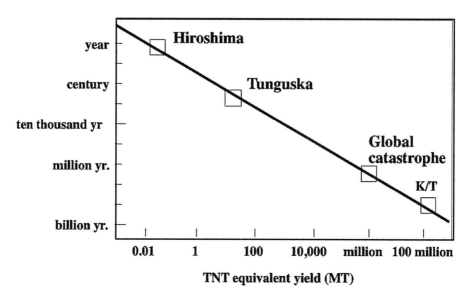

Figure 4.7: The log-log plot of the energy released by an asteroid impact (horizontal axis) versus the frequency of occurrence (vertical axis). The bigger the bang, the smaller the frequency of occurrence. (Illustration courtesy of David Morrison.)

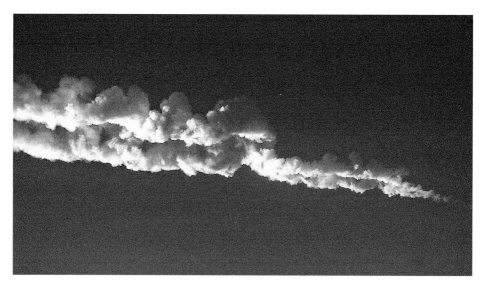

Figure 4.8: Faster than a speeding bullet and as big as a four-story house, the Chelyabinsk meteoroid pays a visit to planet Earth. (Image courtesy of Nikita Plekhanov, Wiki Commons.)

asteroid some 10 km in diameter that struck the Earth at Chicxulub in the Gulf of Mexico. It played a key role in the demise of the dinosaurs and a similar one now would destroy civilization as we know it. Such large events are rare though and happen on the order of every 100 million years or so (labelled K/T in Fig. 4.7). That we are here to discuss this matter at all is simply by the grace of an enormous random asteroid impact event.

Another interesting event again happened over Russia, this time in the Siberian region of Sikhote-Alin in 1947, within the lifetime of some readers. Tons of metallic asteroid fragments rained down on the mountainous region and roughly 70 tons of the material survived[12].

Besides being involved in calculating the end of the world from asteroid impacts, power laws have many other applications, such as in the study of networks like the internet, or when considering phase transitions, as when a solid turns into a liquid or a liquid into a gas. It is also used in Economics to describe the distribution of wealth, whereby a small fraction of a population holds a large amount of wealth, with the bulk of the population holding little. This imbalance has been formalized into what is known as the *Pareto Principle*, better known as the '80/20 Rule'.

Instead of dwelling on how the world could suddenly end, let us explore how everything is slowly unravelling all on its own.

THE WORLD OF ENTROPY – ORDER TO CHAOS

This is the way the world ends / not with a bang but a whimper...
T.S. Eliot

Have you ever noticed how, shortly after you have completely tidied up the house, it soon becomes messy again? Well, there is a name for this tendency of order dissipating to disorder, and it is called *entropy*. This is a concept that has varying definitions depending on the context, from the state of your bedroom to subjects like thermodynamics and information theory, to the ultimate fate of the Universe. In its simplest manifestation, a high degree of entropy is associated with a high state of

---

[12] Some asteroids partially melted and differentiated, like the Earth, forming an iron core and silicate mantle. These bodies were small enough that the core cooled and solidified, and once they broke up in collision with other asteroids, the metal core was released into space. The author has a piece of this meteorite.

disorder, and conversely low entropy means a more organized state and low level of disorder. It is associated with the famous Second Law of Thermodynamics:

Over time, the total entropy of an isolated system never decreases.

In other words: *entropy can only increase or remain the same*[13].

The entire Universe is considered a closed system, since there is nothing acting upon it from the 'outside' (whatever 'outside' means). This leads to the fact that the entropy of the Universe is increasing, and everything is tending towards disorder and dissolution, just like your bedroom only on a grander scale. This consequence of the Second Law is known as the 'heat death' of the Universe, an idea that goes back to English scientist William Thomson (Lord Kelvin).

The poet T.S. Eliot, as quoted above, was right. Sorry to be the bearer of bad news, but the good news is that it will not occur for billions of years yet, and long after our Sun has become a red giant and roasted the Earth anyway (more bad news).

The notion of increasing entropy led astrophysicist Sir Arthur Eddington to propose the notion of an 'arrow of time' in the 1920s, in which the unfolding of events is not time symmetric, but rather time flows in the direction dictated by the Second Law of Thermodynamics.

One might justly ask how things like galaxies, stars, and solar systems form if the tendency of the Universe is towards increasing entropy. In the case of stars, for instance, large molecular clouds begin to collapse due to gravity. If you are having trouble associating clouds of gas with gravity, let us first consider an ordinary cloud in the sky. Taking one that has its length, height, and depth as one kilometer for each dimension, a quick calculation shows that it weighs 500,000 kilograms, or approximately the weight of 100 elephants[14].

So, gas clouds can have tremendous mass, especially when they stretch across vast reaches of space. As the cloud begins to collapse and is confined to an ever-decreasing volume, its entropy actually decreases. But the contracting cloud also generates heat that is emitted as thermal radiation into its surroundings, which are at a lower temperature. This increases the entropy of the Universe as a whole as it more than compensates for the local decrease due to the cloud of gas occupying a smaller volume.

---

[13] There are various expressions of the Second Law of Thermodynamics and the one given above is arguably the most common.

[14] A cumulus cloud typically has a density of $0.5 gm/m^3$. Our cloud of $1\ km^3$ is $10^9\ m^3$ in size, so $0.5\ gm/m^3 \times 10^9 m^3 = 5 \times 10^8\ gm = 5 \times 10^5\ kg$. We are taking the average weight of an elephant in rough terms as $5000\ kg$.

Let us consider a more familiar example. If you take an ice cube out of the freezer and put it on the table at room temperature, you can watch its low entropy ordered crystal structure (Fig. 4.9) change to the high entropy disordered state of flowing water as the ice cube melts. Thus, the entropy of the (effectively) isolated system of the room + ice cube has increased.

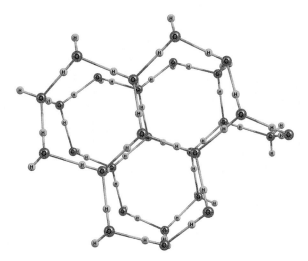

Figure 4.9: The low entropy hexagonal crystal lattice structure of water ice, with each molecule bonded to four others. In its liquid form, the high entropy water molecules are randomly arrayed.

The traditional mathematical symbol to denote entropy is S and both it and the name are due to German mathematician and physicist Rudolf Clausius (1822–1888), who was trying to make the notion of entropy in the Second Law mathematically rigorous. Clausius defined the *change* in entropy, written as, ΔS, when a quantity of heat, Q, is transferred to a body already at a temperature T. This increase is given by the simple formula

$$\Delta S = \frac{Q}{T}. \tag{15}$$

One might think that heat and temperature are really the same thing, but they are slightly different concepts. Heat in a physical sense is a *quantity* of thermal energy. It may be transferred from one system to another and is measured in units called *joules*[15]. Whereas temperature is the measure of the system's internal energy due to the motion of its particle constituents.

---

[15]This unit is related to the notion of *work*, namely the amount of work that is required to produce one watt of power for one second.

Temperature is commonly measured in degrees Centigrade or Fahrenheit, but for entropy the Kelvin scale is used[16]. So in our example of the ice cube on the table, the room (and table) are transferring heat energy to the ice cube, raising its temperature.

This energy transfer – and hence entropy increase – can actually be quantified if we know the temperatures of the ice cube and the room. If the ice cube is at an initial temperature T_1, and that of the room and table are at temperature T_2, then by eq. (15) each will have respective entropy changes:

$$Q / T_1 \text{ and } -Q / T_2 .$$

The minus sign in the second term is because the heat Q is being transferred into the ice cube and out of the room. The net entropy change of the isolated room + ice cube is therefore,

$$\Delta S = \left(Q / T_1 - Q / T_2 \right) = Q \left(\frac{1}{T_1} - \frac{1}{T_2} \right),$$

and the units are joules/kelvin (J/K).

Note that $\Delta S > 0$, since $T_1 < T_2$; that is, the entropy of our closed system of room + ice cube has increased since the ice has melted. If you live in a freezing cold climate and do not turn on the heat, so that $T_1 = T_2$, there is no transfer of heat, i.e. $Q = 0$, so the ice cube does not melt, and consequently $\Delta S = 0$.

A succinct representation of the Second Law of Thermodynamics is that

$$\Delta S \geq 0,$$

for an isolated system.

INFORMATION ENTROPY

Information is information; it is neither matter nor energy…
Mathematician Norbert Wiener

Another important type of entropy concerns information. It was mathematician John von Neumann who suggested to American mathematician and scientist Claude Shannon, the founder of information theory, that he should call his

---

[16] The Kelvin scale starts at absolute zero, which is basically considered to be the lowest possible temperature at $-273°C$. Kelvin and Centigrade are thus related by the formula: Kelvin = $°C + 273$, so that room temperature of $20°C$ is equal to 293 K.

measure of information, *entropy*, since "no one knows what entropy is, so in a debate you will always have the advantage." Shannon took his advice, in his landmark 1948 paper, *A Mathematical Theory of Communication*. Shannon's idea of the information content, *I*, of an event is based on the probability, *p*, of the event happening, namely[17]:

$$I = \log_2\left(1/p\right).$$

For instance, computers handle information in strings of 0s and 1s, each having a probability of 1/2 of occurring, and so the information content of each digit is: $I = \log_2(2) = 1$, and the units are called *bits*[18]. Thus, each digit carries 1 bit of information. Now you know.

If the event is very unlikely to occur – that is, it has a high degree of uncertainty, say one chance in 64 (that is, $p = 1/64$) – then, $I = \log_2(64) = 6$ bits, which illustrates the point that the more uncertainty regarding an event's occurrence, the higher its information content (entropy).

On the other hand, if the event is very likely to occur – that is, it has low uncertainty, say the probability of rain tomorrow is 90% – then the information content is: $I = \log_2(10/9) = 0.152$ bits, which means that the information held few surprises for all those knowing souls carrying an umbrella. So, in information theory, entropy is a mathematical measure of uncertainty. More uncertainty equals more entropy, less uncertainty equals less entropy.

This works out nicely for if $p = 1$ – that is, an event is *certain* to occur – then,

$$I = \log_2\left(1\right) = 0,$$

meaning no information is conveyed.

Computers use electricity to process information, with the aid of electronic switches called *transistors* that can now be made so small that billions of them can fit onto a fingernail-size computer chip. The transistors control the flow of electricity and can be in one of two different voltage states, often denoted by ON or OFF, respectively. By associating the number 1 with ON and the number 0 with OFF, this allows for computations and data storage to be carried out in terms of nothing more than strings of ones and zeros. See Chapter 7 (Qubits) for a simple example.

If we have *N* switches that can be either ON (1) or OFF (0), then the ensemble can be in any of 2^N different states of 0s and 1s. Thus the probability of being in any one state is $1/2^N$, and consequently,

$$I = \log_2\left(2^N\right) = N,$$

---

[17] The base of the logarithm taken here is 2, but other systems use the number 10, 3, or *e*.

[18] The term 'bits' (from 'binary digits') was suggested by mathematician John Tukey and also used by Shannon.

meaning that N bits of information can be stored by the N switches.

In the case where multiple events can occur, then we take a 'weighted sum', with each weighting given by its respective probability of occurrence to obtain the *average* information entropy of the system. In other words, if there are N independent events, each with probability of occurring: p_1, p_2, ..., p_N, respectively, $\left(\sum_{i=1}^{N} p_i = 1 \right)$,[19] then the *information entropy* is given by the sum

$$I = \sum_{i=1}^{N} p_i \log_2 \left(1/p_i \right).$$

This formulation of entropy probably goes back to John von Neumann (1927) and his work on statistical mechanics, which is the study of large ensembles such as gases.

One thing Shannon was interested in was the average information gained when we learn one letter of the English alphabet from a given sentence. As an experiment, subjects are placed before a sequence of dashes, each representing a letter of the (English) alphabet or a space. These comprise a particular English sentence and the subjects have to guess successively what each dash represents. This is best done on a computer and when the subject guesses correctly, the computer replaces the dash with the correct letter (or space). Each decision by a participant is based only on the information gained from previous correct guesses.

The number of guesses needed to hit upon the correct letter/space are recorded in each instance. From this, an average can be computed that tells how much information is gained when we learn a correct letter. Experiments of this nature have yielded an average value of 1.1 bits of information gained from knowing a letter of a sentence[20].

There are many aspects to entropy, and we have only touched upon a few. Indeed, in a recent study, nearly 140,000 paintings were analyzed for their degree of entropy, and as you might suspect, Minimalist paintings exhibited small entropy values, whereas works such as Jackson's Pollock's drip paintings (Fig. 4.10) had large entropy values[21].

One interesting feature is that it is possible to reduce entropy with energy and information from an outside source. This is how living organisms such as plants and animals can organize themselves into exquisitely complex structures. They are not closed systems but open to the energy of the Sun and to their natural environment.

---

[19] Here, we use the short-hand notation for writing a sum: $\sum_{i=1}^{N} p_i = p_1 + p_2 + \ldots + p_N$. The Greek letter Σ is the uppercase sigma and in Mathematics generally stands for the sum.

[20] An applet for this experiment is available at: http://www.personal.psu.edu/dpl14/java/informationtheory/entropy/index.html

[21] H.Y. D. Sigaki, *et al.*, History of art paintings through the lens of entropy and complexity, *Proc. Nat. Acad. Sci.*, 115, E8585–E8594, 2018.

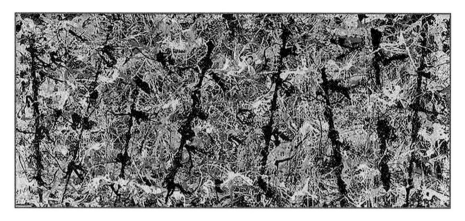

Figure 4.10: The high-entropy painting Blue Poles by Jackson Pollock. (Image in public domain.)

Moreover, information becomes encoded in our DNA in the form of instructions to make the proteins necessary for life. This information has the effect of reducing entropy in living organisms.

But with the passage of time, this process of keeping entropy at bay becomes less and less efficient, and our momentary victories over entropy become unsustainable.

Nor does Darwinian evolution defy the Second Law, as this requires the outside intervention of the forces of natural selection.

The study of entropy has received renewed interest in recent years because its underlying nature is rather mysterious. The preceding comment by von Neumann, that no one knows what entropy is, was not really made in jest. A new understanding of entropy is being sought in the field of Quantum Mechanics that is based on information garnered from measurements taken of a quantum system. So, a very 19th century notion is now undergoing a 21st century make-over which may shed some light on the real meaning of this strange word, entropy[22].

In the next chapter we will see how sometimes things can work in reverse, and that order can manifest in some systems from chaos and not even necessarily by design.

---

[22] M. Brooks, Driver of Disorder, *New Scientist*, 7 Dec. 2019, 34–37.

5

Order from Chaos

I have often made the hypothesis that ultimately physics will not require a mathematical statement, that in the end the machinery will be revealed, and the laws will turn out to be simple, like the chequer board with all its apparent complexities...
Nobel Prize winning physicist Richard Feynman

One question of great interest to people who think about the complexity that is found everywhere in Nature, is how did it get there? Somehow, Nature manages to reverse the tendency to disorder as embraced in the Second Law of Thermodynamics. So, let us examine a few unusual ways to reverse the process and obtain complexity rather than disorder.

CELLULAR AUTOMATA

In what follows, a very simple mechanism will be demonstrated that uses the most primitive tools, yet is able to achieve the most remarkable degree of complexity. The basic framework involves no mathematical formulations − although Mathematics comes into play when creating actual simulations of natural processes − and can yield quantitative as well as qualitative results.

Even the way this complexity arises is not really transparent, it just happens. Yet the rules of engagement are simplicity themselves. For those wishing to pursue the matter further, see the author's book on Cellular Automata (CA) in the Bibliography. Everything that that follows here about CA is more or less contained in that text, but is presented there on a more technical level.

© Springer Nature Switzerland AG 2020
J. L. Schiff, *The Mathematical Universe*, Springer Praxis Books,
https://doi.org/10.1007/978-3-030-50649-0_5

The notion of a cellular automaton came from the outline of an idea originating with Hungarian-American mathematician John von Neumann, regarding a machine that could replicate itself. This led to the creation of a CA environment by mathematician Stanisław Ulam who, with colleague R.G. Schrandt at the Los Alamos National Laboratory in New Mexico, created various CA computer simulations during the 1960s. Subsequently, John Conway's CA-based *Game of Life* distracted government employees for countless hours as it played out on computers in the 1970s and 80s. It still remains a distraction for some.

It was only in the 1980s that Stephen Wolfram seriously began to investigate the properties and utility of CA, and research continues to the present day with applications to be found in numerous scientific fields, including Biology, Physics, and Astronomy. Indeed, it has even been suggested, most notably by computer scientist Edward Fredkin, that all of Nature (including space and time) is discrete and that the entire Universe is one giant cellular automaton[1].

A few words must be said about the great genius who started all of this, John von Neumann (1903−1957), one of the giants of 20th century Mathematics and Physics. Von Neumann was a child prodigy with uncanny mental prowess, who worked with Edward Teller and Stan Ulam on the Manhattan Project during World War II. He is the creator of the important field known as Game Theory and, with German economist Oskar Morgenstern (1902−1977), wrote the seminal text on the subject[2].

Game Theory is a much broader field than its name would suggest, being a crucial component in Economics, conflict analysis, psychology of human interactions, Politics, and even Biology, where it is known as Evolutionary Game Theory and models the struggle of populations for survival. A Darwinian model will be given later in the text which derives largely from the Theory of Evolution applied to computer algorithms.

In 1932, von Neumann laid the foundations for the rigorous mathematical formulation of Quantum Mechanics (see Chapter 7), and in 1945 he mapped out the framework for modern computer architecture. There are so many fields in which von Neumann made substantial contributions that it would take another book just to introduce the topics.

So, let us humbly go back to von Neumann's 1948 proposal for a self-replicating machine and the CA universe suggested by Ulam. Suppose you stub your big toe. The cells in the immediate vicinity of the affected area will then spring into action, but the cells of your hands and arms will be oblivious to the damaged toe, and so the repair to the injury can be considered to be a 'local' phenomenon. Or suppose

---

[1] E. Fredkin, A new cosmogony, *Proc. of the Workshop on Physics and Computation*, D. Matzke, ed., IEEE Comp. Sci. Press, New York, 1993, pp. 116−121.

[2] J. von Neumann, and O. Morgenstern, *The Theory of Games and Economic Behavior*, Princeton University Press, 1944.

you drop a pellet of cyanide into a reservoir. It will have no effect on the water far removed from the drop site, at least not for a very long while (this is the process of diffusion). In fact, if you think about many natural phenomena, they tend to be local in nature.

This local nature of such phenomena is the basis for constructing cellular automaton models. Although they are, of necessity, highly simplified models with only three guiding principles, they have been very successfully engaged in modeling all sorts of natural phenomena. They can be used to reproduce the patterns found on many species of seashells, in modeling the growth of snow crystals, in models of how neighborhoods become segregated, or how bacteria colonies spread, and much, much more. This is what the above-mentioned text explores in greater detail.

For this reason, it is also an important tool for scientific investigation into the nature of how the world works. Cellular Automata can be simulated in 1, 2, 3, and more dimensions, but we will only discuss the first two dimensions here since that will serve all our purposes. Contrary to the conventional approach, let us start with two dimensions and examine the simpler one-dimensional case afterwards.

We are now going to create our 'universe' (with a very small 'u'), with the squares of a checkerboard grid, or array. Each square can be thought of as a real 'cell' like a patchwork of skin cells, as in Fig. 5.1.

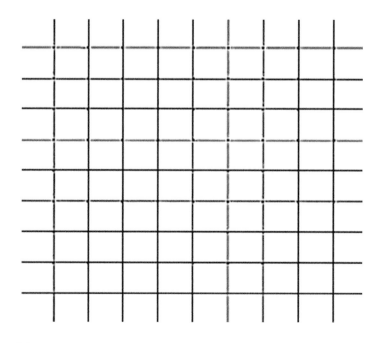

Figure 5.1: The grid of identical cells that form the landscape of the two-dimensional Cellular Automata (CA) universe.

The size of each cell is of no importance, and the extent of the array can be thought to be indefinite at this stage. Later, we will mention the possibility of imposing boundaries on our array universe.

Let us now assume that each such cell can exist in one of two *states*: either 'dead' or 'alive', just like in real life; either you are dead or alive, hopefully the latter. In computer-speak, 'dead' = 0 or OFF, 'alive' = 1 or ON. This is just like the workings of transistors in a computer, representing the numbers 0 or 1, but now we are just dealing with square cells in a plane, not electronic switches.

If the cell is alive, then to indicate this fact the square will be colored black, and if the cell is dead then the cell will be white in color. Fig. 5.2 is a CA grid (with lines suppressed) with various alive and dead cells. The choice of the two colors is optional but black and white are the ones used traditionally.

Figure 5.2: A cellular automaton with a configuration of live cells (black) and the remaining dead cells (white). Some live (black) cells were initially chosen at random on our two-dimensional grid and the resulting structure seen here is from iterating the Game of Life for several time-steps, as discussed later in the chapter. The grid lines have been suppressed.

This is all simple enough. Now all we need is some way of keeping time as our universe is going to evolve, since a static universe is of no interest. In order to keep time, and again to keep matters as simple as possible, let us add an imaginary clock to our setup, hovering in the background, which makes a tick at every passing second. *It is the ticks that we are concerned with, not the interval between them.* What this says is that 'time' is discrete, 0, 1, 2, 3, …. Using discrete time is useful in treating many phenomena, not just in CA. Digital signal processing is done by taking samples of the signal at discrete times, giving us digital TV, cell-phones, and music. Moreover, in the Loop Quantum Gravity model mentioned in Chapter 8, time *is* discrete along with space.

Each such tick of our clock can be thought of as a 'generation' and our system will evolve through many generations, beginning with some initial state (when the time = 0). The tick-tock of time does not have to occur in actual real-time seconds of course, but merely keeps the cadence of the updating regime.

Okay, that is our 'universe' and as we are playing the role of a Creator, we now need some basic guiding principles. There are only three. These conditions will apply to how the cells in our universe are allowed to change their state, according to some yet to be specified 'Rules'. We will make up some interesting Rules as we go along, as we are the Creator, remember. The Rules will instruct each cell at each tick of the clock to change its state from either dead to alive, or alive to dead, in a simple analogy with birth and death. But first we need the three guiding principles for the overall functioning of our universe.

Here are the CA Principles:

i. **All cell states (alive or dead) are updated by the same set of Rules.**
ii. **All cell states are updated simultaneously.**
iii. **The Rules are local in nature.**

Regarding the first *Uniformity Principle*, we cannot use one set of Rules on some cells and another set of Rules on other cells. Every single cell in our array must abide by the same set of Rules, whatever they may be.

The second *Principle of Synchronicity* means that at each tick of the clock, all cells change their state according to the Rules at the same instant, which we regard as instantaneous.

The final *Principle of Locality* needs some explanation as to what we mean by 'local'. Notice that in the following Fig. 5.3, the central cell on which we fix our attention for the moment has eight adjacent neighboring cells that surround it[3].

These cells will be the *only ones* that we will consider to be in the local neighborhood of the central one, and the only cells that affect the state of the central cell. Of course, every cell in our grid array has a neighborhood consisting of eight such cells. No cell is an island.

In order to set our universe in motion, we simply need a Rule, one that determines the state of every cell at the next generation. This will be done by examining the state of all the neighborhood cells at the present generation, in conjunction with the present state of the central cell.

---

[3] Another neighborhood used in CA considerations is the *von Neumann* neighborhood, consisting of the four cells that lie *N, S, E, W* of the central cell. The eight-cell neighborhood that we are using is a *Moore* neighborhood, named for Edward F. Moore, an American computer scientist.

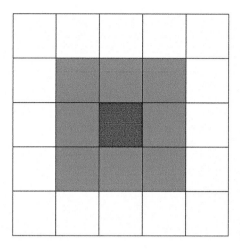

Figure 5.3: An eight-cell (blue) neighborhood that is used in CA to update the state of the central cell, based on the states of its neighboring cells and its own current state. (Illustration courtesy of Katy Metcalf.)

In general, when using a finite-size grid to fit on a computer screen, we need to have some convention to deal with the cells at the boundary of the grid. One common option is to make the cells at the top and bottom neighbors of each other by mentally folding the screen into a cylinder. This is more easily visualized with a sheet of paper. The cells on the left and right sides are also made neighbors of each other. This effectively turns the flat screen into a donut (torus) shape and thus no cell is left out of the action.

However, there will be certain occasions when we want to fix the values of the cells on the boundary, such as when computing the temperature inside a region (as in Fig. 5.10), and in this instance the boundary cells do not need any neighbors.

Here is a simple illustration of a CA Rule, starting from a random initial configuration of 1s and 0s in our array. At each time-step:

i. *Compute the sum of the state values for each cell plus its eight neighbors.*
ii. *If the sum comes to 5 or more (a majority: 5 of 9), then the central cell takes the value 1 at the next time-step:*
iii. *If the sum is 4 or less (a minority), the central cell takes the value 0 at the next time-step.*

This is an actual CA Rule, and in fact it has a name: *Vote* (or *Majority*).

In anthropomorphic terms, we could consider the 1s and 0s as representing political parties, say either Democrat or Republican. Of course, we are assuming that every person lives in a cell (on their own) of a grid-like array, with every neighbor in either one party or the other. A gross simplification without a doubt, but useful (also without a doubt). In the *Vote* scenario, as we let the system evolve, we find that we end up with static pockets of voters of like-minded political persuasion (Fig. 5.4).

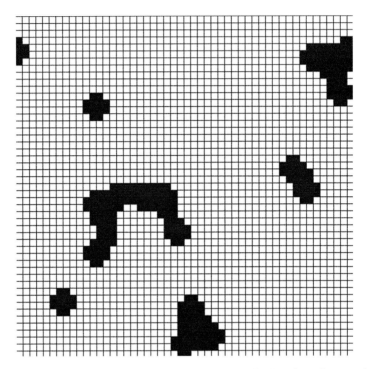

Figure 5.4: The CA model *Vote*, whereby the central cell takes the value 1 at the next time-step if the state values of the cell and its eight neighbors sum to 5 or more. Otherwise, the central cell takes the state value 0. Starting with a random configuration of live and dead cells, the result is static pockets of cells having the same state value. The Left/Right and Top/Bottom edges have been joined in the evolution of the model.

A related model derived by American economist Thomas Schelling involved the propensity of people (simulated as cells of a grid-like array) to move from their neighborhood if too many of the neighbors were not of the same state, i.e. either black or white on the array, as they were. Schelling found that even with a fairly high level of tolerance shared by all 'people cells' for those of a different state, the array still evolved into numerous segregated pockets, each consisting of just a single state, either all black or all white. This is a somewhat similar phenomenon to the *Vote* scenario. In part, for this work, Schelling shared the 2005 Nobel Prize in Economics.

LIFE AS A GAME

We are now going to introduce some very simple Rules that come from what is known as the *Game of Life*, which was created by distinguished English mathematician John Horton Conway (1937–2020) and his students at Cambridge University, and famously published in the October 1970 issue of *Scientific American*:

   i. *A dead cell (white) becomes alive at the next time-step (generation) if exactly three of its eight neighbors are alive, otherwise it remains dead.*

  ii. *A live cell (black) at the next generation remains alive if either two or three of its eight neighbors are alive, but if more than three are alive or less than one is alive, then the cell dies.*

Those are the rules of the Game of Life. Conway was trying to get the system of cells in some vague sense to encapsulate the essence of a living system, as well as being sustainable for a long sequence of time-steps.

Rule (i) shows how life is created and it takes three neighbors to tango in Conway's world. So, if a cell is dead but has exactly three live neighbors, then it will itself blossom into life at the next tick of the clock. Rule (ii) reflects the notion that if a live cell has too many live neighbors – more than three − then the cell is going to die, say, due to 'overcrowding', and if it has too few live neighbors – only one − then it will die from loneliness or boredom. Bear in mind that this is just a friendly example of a cellular automaton and is in no way a model of real life or any life processes.

When the Game of Life first came out it caused an immediate sensation, and computer screens across the globe were running simulations of it. Of course, it is not really a game at all, but the one thing that you can alter is the initial arrangement of live and dead cells across the array. Then you have a computer program which implements the two Rules to update every cell at each new generation.

Originally, Conway and his students used a *Go* board with black and white stones to indicate the state of each cell, but that is far too tedious in the present computer age and there are several implementations that are available on the internet[4]. All you have to do is indicate an initial configuration of live cells (the remaining ones are dead by default), and the computer sequentially goes through the updating of the array at each subsequent generation.

It helps your thinking if you view each cell of the array as a mini-computer. Firstly, at what would be time = 0, we start with some initial configuration of alive and dead cells (black and white respectively). At each subsequent tick of the clock, the cell looks around and examines the state of each of its eight neighbors. The

---

[4]For example, the website, https://bitstorm.org/gameoflife/standalone/ has downloadable Windows, Mac, and Linux versions.

cell then applies the Rules (i) and (ii) to see how it should update its current state, from alive to dead, or dead to alive, or simply remain the same. Every cell of the array goes through this very same process simultaneously, and after a number of generations we, the Creator, can stop the whole process and peer at the results. That, essentially, is the Game of Life.

Let us try it. Here are some practice runs involving three live cells (Fig. 5.5), each of which comes to a halt fairly quickly. The author has made the initial setup quite simple so that, if you wish, you can verify that the Rules (i) and (ii) are being followed at each generation.

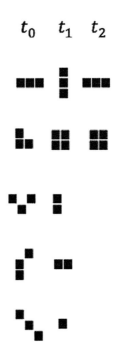

Figure 5.5: The evolution of five initial configurations on the left after two subsequent time-steps (to the right), in accordance with the rules of the Game of Life. In the last three examples, no live cells remain by the second generation – life has been extinguished. Grid lines have been suppressed.

Notice that a single live cell will die (Rule (ii) – loneliness) and all dead cells will remain dead if all its neighbors are dead (Rule (i) – no sex). Even two adjacent live cells will become extinct at the next generation for the same reasons. We really need at least three live cells in proximity to each other to get things going properly. In fact, if we have three live cells in a row, as in Fig. 5.5 (top), we see that by Rules (i) and (ii), the first generation becomes three vertical live cells, which at the next generation become horizontal again, and which at the next generation become vertical again and so on. Thus, there are two different configurations that the cellular automaton oscillates between, and we call this a 'period two oscillator', a notion that we will see again when discussing Dynamical Systems.

It turns out that numerous other period two oscillators have been discovered by clever arrangement of the live cells in the initial configuration. Moreover, oscillators of various other periods are known, some with high periods such as a 29-cell 'glider shuttle' of period 60, found by David Buckingham.

The Game of Life has been extensively studied since its initial appearance on the world stage in 1970, and many remarkable features have been discovered in it. One of the most intriguing is that of a certain type of gun that fires projectiles. The gun is called a 'glider gun', as the projectiles are known as 'gliders'. In Fig. 5.6, we see the glider gun in action. It was created by Bill Gosper of MIT, winning a $50 prize from Conway for his efforts as it demonstrated that a finite number of live cells in the Game of Life could grow *ad infinitum* over time.

Figure 5.6: A glider gun (top configuration) and three glider projectiles that it has manufactured and shot out. The gliders will move at each time-step in an ungainly manner to the right along the diagonal. Grid lines have been suppressed.

Now let us take this collection of live cells that compose the glider gun in an array and start the clock ticking. As the generations tick over, the live and dead cells come and go, until the 30th generation. At that particular moment, a projectile is produced – the glider. What the glider then does in subsequent generations is move diagonally one cell at a time (i.e. one cell per time-step). The motion of the glider is really something to behold, as it looks very much like a living organism with an awkward gait. It looks uncannily alive as it edges across the screen and is fascinating to watch, but its every movement is controlled entirely by Rules (i) and (ii), nothing more.

Meanwhile, as the first glider is making its ungainly way diagonally across the screen, the glider gun is gearing up to fire off a second glider at generation 60. The process goes on, generation after generation repeating the same cycle of firing off another glider projectile every 30 time-steps.

Besides the fascinating activity that is unfolding on the computer screen, there is something much deeper going on here. The firing of a glider can be viewed as the sending of a pulse of information to a distant location.

As it turns out, it has been demonstrated that by using nothing more than the Game of Life Rules, arrayed with suitably arranged glider guns and suitable receptors, it is possible to create *logic gates* that provide the necessary circuitry for a computer to function (see Appendix XIII)[5]. It would, in principle, be as powerful as any general-purpose computer in the world. The words 'in principle' have been added because it would actually be incredibly cumbersome to construct such a computer, but the fact remains that it could be done. This extraordinary ability of the Game of Life to function as a computer is called 'universal computation' and is a rather remarkable consequence of what, on the surface, appears to be a very primitive affair.

So, let us summarize what we have just done. We started out with the three guiding principles that govern the evolution of any cellular automaton, namely Uniformity, Synchronicity, and Locality. These are the basic requirements for any array of cells if they are to become a cellular automaton. The next ingredient we needed was a set of Rules, and in this case, we only required two, the simple Rules (i) and (ii) governing cell birth and cell death. Then, once we are given some initial configuration of cells that are alive in our grid universe, we start the clock ticking, 1, 2, 3, …, and observe the generations of cells doing their dance of life and death unfolding before our eyes.

Furthermore, as mentioned above, it is found that the basic Rules (i) and (ii) have the remarkable power (in principle) of a general-purpose computer through their ability to create logic gates, thus opening up the entire world of computation.

The reader is urged to download the Game of Life applet noted in Footnote 4 above and simply put in some initial live cells and let it run. You will be amazed at how life-like the whole scene looks, with some structures running into others and annihilating them or reforming into other ones, with gliders accidentally produced at times that smash into other objects only to disappear, and so forth. Within such a simple universe, we find the seeds of 'artificial life', which has become a subject of study in its own right.

One final note before we move on. Because the computer screen can only be made so large, a boundary on the array of cells is usually imposed. In principle, an unopposed glider would simply go off to infinity (wherever that is), but a computer screen only has a finite size. So as mentioned previously, we make the top and bottom borders with each other and likewise with the left and right sides. This feature keeps the action going and then no structures can simply wander off the screen and never be seen again. They will simply reappear on the opposite side of the screen. You will see this feature in some implementations of the Game of Life on the internet.

---

[5] Logic gates such as AND, OR, and NOT are electronic circuits that govern the flow of electricity from one or more inputs to a single output. They derive from symbolic logic that was touched upon in Chapter 1, and how they work is discussed in Appendix XIII.

INFECTIOUS DISEASE MODEL – SIR

Let us now consider a more serious application of CA, namely the SIR model of infectious disease. We will present here only the most basic elements of the model, as there are many more sophisticated versions of it in the literature that are used by epidemiologists. This model is especially pertinent given the pandemic of the Covid-19 virus.

In the model, we consider the eight-cell Moore neighborhood as in Fig. 5.3. Each cell can be in one of ten states:

S: susceptible (state 0 – *white*)
I: infectious (states 1, 2, 3, 4, 5, 6, 7 – *shades of red*)
R: recovered (state 8 – *green*)

We will add a further cell state, 9 (black), meaning 'deceased'. As with many infectious diseases, not everyone will survive, so let us say 1 in 100 infected cells will not make the transition to the recovered state 8 but will go from infectious state 7 to deceased state 9.

The Rules for the update of the state of each cell are as follows

  i. *A susceptible cell (individual) in state 0:*

   ▪ *will transition to the infectious state 1 at the next time-step if two or more of its neighbors are infectious*
   ▪ *otherwise it will remain in state 0 at the next time-step.*

 ii. *An infectious cell (individual) in state k, for k = 1, 2, 3, 4, 5, 6, will transition at the next time-step to state k + 1.*

iii. *For an infectious cell that has reached state 7:*

   ▪ *99% will transition to state 8 (recovered) at the next time-step*
   ▪ *the remaining 1% will transition to state 9 (deceased) at the next time-step.*

 iv. *A recovered cell (individual) in state 8 will continue to remain in that state.*

Here, we have taken the threshold for an individual cell to become infected as having two or more infected neighbors. It is more realistic to say that an individual cell would become infected with a certain probability depending on the number of infected neighbors, and this is what more sophisticated models do. In other words, the more neighbors who are infected the higher the likelihood of the central cell becoming infected.

Note that once a cell becomes infectious it remains so for a period of seven time-steps, which one can think of as seven days. After seven days of infection, 99% of those cells transition to the recovered state the next day and 1% will die (Fig. 5.7), which is very roughly the mortality rate of the Covid-19 virus. The number of infectious days was chosen for illustration purposes only and of course the version of the SIR model presented here is very simplistic. The actual dynam-ics of disease contagion are far more complex than has been illustrated here, but so are the mathematical models.

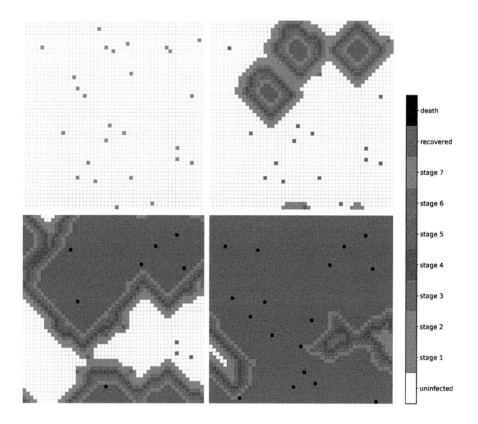

Figure 5.7: The spread of a virus from an initial random sprinkling of infected cells (red – first panel) among a population of susceptible but healthy white cells according to the SIR model in the text. The seven infectious states are indicated by a shade of red: light coral, coral, tomato, crimson, red, tomato, light coral, respectively. The second panel is after 10 iterations, while the bottom left panel is at the height of the epidemic after 20 iterations, where many individuals have now recovered (green) and some have died (black). As the epidemic subsides in the bottom right-hand panel after 30 iterations, only isolated pockets of infected cells persist. Eventually, only green and black cells will remain. Quantitative data can be extracted from the model by comparing white, red, green, and black cell numbers at different stages. (Image courtesy of Anita Kean.)

MIMICKING DARWIN

It is very worthwhile to mention a further CA application because it is very germane to the subject of order from chaos, and that is the notion of 'genetic algorithms.' Just from the name, it is clear that this combines features of Biology with Mathematics.

In particular, the biologic features that are utilized are those of evolution itself, specifically 'natural selection'. The question now is how it can be applied to solving mathematical problems. Here is where the other word, 'algorithm', comes into play.

Firstly, the author would point out that the word algorithm, which is used a lot in the mathematical sciences, is really just a method for solving a problem involving some sequence of operations. The word itself derives from the notable Persian mathematician Muhammad bin Mūsā al-Khwārizmī (*ca.* 780−850), whose work became the foundations of Algebra.

In essence, we are faced with a particular problem; for example, we have a wholly arbitrary placement of black and white (alive and dead) cells in our two-dimensional universe. They are in a completely random jumble. The problem before us is to find some general procedure (algorithm), obeying the general principles of our CA universe, that will eventually transform the jumbled cellular array (or any such random array) into the pattern of a checkerboard, with alternating black and white cells. Order from chaos.

Just an outline of the procedure will be given, as the full account containing all the details can be found in the author's book on Cellular Automata (see the Bibliography).

Firstly, we begin with a random pool of algorithms that serve merely as a starting point. Pure guesswork sometimes has a respectable place in Mathematics, generally serving as a starting point such as in the famous Newton-Raphson method of finding the zeros of a function. So, we are on reasonable grounds here. Now, armed with this pool of algorithms, we can try them out on the problem at hand, i.e., changing the states of the randomized cells into an orderly checkerboard.

We then have some method of measuring their performance − that is, a 'fitness test' − and with a big enough starting pool of algorithms, some will naturally perform better than others. The algorithms that perform below average on this fitness test are weeded out − survival of the fittest.

Of course, the author is leaving out the technical details, but suffice to say that the algorithms basically involve nothing more than strings of 1s and 0s that happen to be 512 digits long (nothing for a computer), and can be thought of as 'chromosomes'[6]. The remaining chromosomes which show some degree of promise in ordering our jumble are then given some special treatment, to try and enhance their performance still further. We can do this in three ways:

- By *cloning*, i.e., taking multiple copies of chromosomes according to how well they performed on the fitness test. The highest achievers will be cloned the most and the lowest achievers cloned the least.

---

[6]The number 512 comes from the fact that the central cell plus its eight neighbors, that is nine cells in total, can be in $2^9 = 512$ different states (configurations) of black and white cells. The value of the central cell at the next time step for each of these *neighborhood-states* can be represented by a string of 0s and 1s that is 512 digits long.

- By *breeding*, i.e., selecting a digit at each gene position at random from one of two parent chromosomes to produce a 'baby' chromosome, also 512 digits long.
- By *mutating* a few of the digits on a chromosome to add some genetic variation to the pool.

Actually, the chromosomes that breed are having a form of digital sex, by having each parent chromosome give up a random digit (a 'gene') and thus producing a 'baby' chromosome which is then added to the pool. Who says that Mathematics is not sexy?

The mutation step was most effectively achieved by randomly flipping one or two digits from either 0 to 1 or vice versa.

We then apply the fitness test to this new pool of adulterated algorithms (chromosomes) and repeat the above procedures. Gradually, with each generation, the algorithms become more and more fit by the standards of the fitness test. In the particular problem at hand, it took the genetic algorithm method only 100 generations to solve the problem and produce order from chaos (Fig. 5.8). The credit for this very clever implementation of a genetic algorithm goes to a former student of the author, Michael Brough, of Auckland University.

Figure 5.8: A genetic algorithm evolution of a random array of black and white squares (top left) into a checkerboard pattern (bottom right). (Image courtesy of John Wiley & Sons, reprinted with permission.)

Other evolutionary approaches to various artificial intelligence problems use less goal-directed and more open-ended search techniques. Such approaches are proving useful in developing algorithms for self-driving cars. This approach is nicely explored in the book by Stanley and Lehman listed in the Bibliography.

Let us consider another two-dimensional application and return to a former example in Chapter 4, where we were interested in determining the (steady-state) temperature distribution inside a circular region. We know by the mean-value property that the temperature at each point is the average of all the temperatures in a circle surrounding the point.

Using our two-dimensional grid to model the region inside the circle, we transform the scenario to that of an array of cells, each with its surrounding eight neighbors (except for the cells on the boundary). As we did previously, we will give the boundary cells on the top half the value $100°$, and the ones on the bottom half $0°$. If the boxes forming the grid are sufficiently small, the outer cells will very reasonably approximate a circle.

We next need to implement the mean value theorem for each cell which has four contiguous neighbors N, S, E, W, and four neighbors that meet at a point, NE, SE, SW, NW (Fig. 5.9). We will give the latter four neighbors one-fourth the influence on the temperature of the central cell compared to the contiguous four neighbors. Thus the 'average' temperature value of the central cell at the next time-step would be:

$$T = \frac{(N+S+E+W) + \frac{1}{4}(NE+SE+SW+NW)}{5}.$$

We divide by 5 as we are considering 5 *whole* values when computing the average for the central cell. Typically, in scientific applications of CA, the model is expressed in mathematical terms, such as in the present case.

With this formula for updating the central cell at each time-step, we let the cellular automaton run its course until it settles down to an equilibrium state. To visualize the results, we can assign colors to the different temperatures − red, orange, yellow, green, blue − to indicate hot to cold (Fig. 5.10). Moreover, we can pick any cell in the grid and query its temperature.

So, for example, let us consider the temperature at the point $(0, 0.5)$, that we looked at previously by taking a random walk. The CA temperature for the box containing this point is, $78.9°$ which compares very favorably with the values of $79.2°$ and $79.3°$ obtained by the random walk method, and the value of $79.5°$ obtained via some serious Mathematics[7].

As a bonus, we have a qualitative image of how the temperature is distributed throughout the enclosed region.

---

[7] This can be done by using what is known as the *Poisson formula*, which gives the solution to the Dirichlet Problem for a circle.

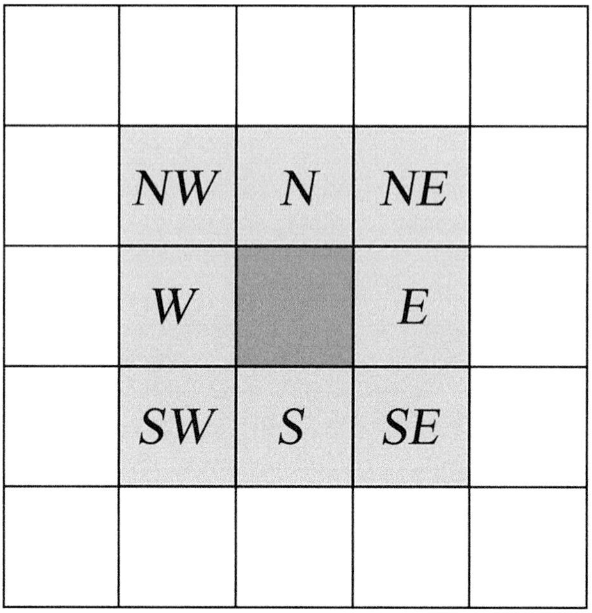

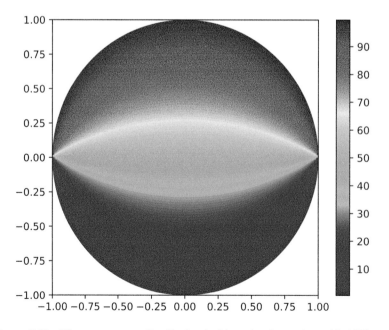

Figure 5.9: The eight neighbors whose temperatures are considered to affect that of the central cell at each subsequent time-step.

Figure 5.10: The temperature distribution inside a circular region with $100°$ on the top half of the boundary and $0°$ on the bottom half. (Image courtesy of Anita Kean.)

ONE-DIMENSIONAL CA

Now we turn our attention to the one-dimensional case of Cellular Automata (CA), which involves an endless (horizontal) row of boxes, each capable of being either alive (black) or dead (white), just as in the two-dimensional case. The same three guiding principles of Uniformity, Synchronicity, and Locality apply, as does the discreteness of time given by our imaginary ticking clock.

For any particular cell, however, there are now only two neighbors, one on the left and one on the right. At any particular time-step, the cell under consideration and its two neighbors can be in exactly $2^3 = 8$ different arrangements of black and white cells, as shown in the top row of Fig. 5.11. These can vary from all white (dead) to all black (alive) plus the six other combinations of dead and alive states for the central cell and its two neighbors.

The Rule governing the behavior of the cellular automaton is determined by what happens to the central cell at the *next* time-step, in the case of each of these eight different configurations. Thus, the behavior of every cell in the one-dimensional line of cells is determined at each time-step, since every cell will find itself in one of the eight configurations with its two neighbors. The updated states for each cell are displayed immediately below the cell, so that the row of cells grows downwards with time, each row representing the update (next generation) of the previous row.

It turns out that in using just two allowable states for each cell, black and white, there are exactly $256 = 2^8$ rules we can apply to our system and that is all. Not many, you might be thinking, so if we were to use a third color, such as gray, in addition to black and white, we would have 7,625,597,484,987 possible rules to choose from.

Or, we could simply re-define what it means to be a neighbor and say that the two adjacent cells to the left and the two to the right of any particular cell are its neighbors. We are in charge of this universe of ours and have a certain amount of freedom when it comes to choosing neighbors. Even in the two-dimensional case, instead of taking all the eight cells that surround a particular cell as the neighborhood, we could have simply taken the four adjacent cells that lie in the North, South, East, and West directions.

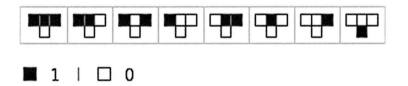

Figure 5.11: The graphic depiction for Rule 1 as described in the text. The eight possible initial configurations are on the top row, with the rule giving the output in each case at the next time-step recorded on the second row. (Image courtesy of Wolfram Alpha.)

But let us be content with our 256 rules as some have truly amazing features. They are actually numbered in order to keep track of them all, from Rule 0 to Rule 255. Let us start with a simple one, namely Rule 1:

The central cell changes to white at the next time-step in every case, except when the central cell and its two neighbors are all white, in which case the central cell changes to black.

The situation can be depicted graphically, giving the eight possible configurations for a central cell with a left-hand and right-hand neighbor, as in Fig. 5.11, with the update according to the rule below the central cell. The rule is unusual in the sense that all three-cell configurations of live and dead cells lead to a dead cell output, except that three dead cells can bring forth a live cell. Of course, we are only looking at one of the simplest of the 256 rules.

Actually, Rule 0 is the simplest of all, as every configuration of three cells leads to a dead cell (0) evolution of its central member, so that rule is not terribly interesting.

Rule 1 is simple indeed, and Fig. 5.12 displays the first 100 time-step evolution of the rule, starting off with a random array of alive and dead cells (black and

Figure 5.12: Note the repetition of various black/white patterns exhibited by Rule 1, even though the initial conditions were a random selection of black and white cells along the top row and the rule is about as simple as it can get. (Image courtesy of Wolfram Alpha.)

white). In order to see the evolution of any rule, we start at time $t = 0$ and must 'seed' the initial row of cells, each with a starting value. This can either be randomly chosen, or simply start with one black cell and all the rest white.

The initial set of states run along the very top row, and the output at each subsequent time-step is represented by the following row, so that time proceeds downwards in accordance with the chosen rule. In principle, the horizontal arrays extend to the left and right indefinitely, but life is finite and so the array depicted is as well.

Again, the cells at the far left and far right only have one neighbor, so to get around this issue we make them neighbors of each other, as was done in the two-dimensional case. Thus, the furthest left cell now has a neighbor to its left, namely the cell to the furthest right, and vice versa.

Starting with an initial seeding of black and white cells that is completely random on the top row, what we observe from Rule 1 is that it has managed to organize its output in a very regular fashion (Fig. 5.12). There was nothing very complex about Rule 1 either, so the partially organized output is somewhat unexpected. We are getting some structure from a near-nothing rule.

Such CA as these belong to the same class of automata that we found in the Game of Life, with patterns that repeated cyclically.

Next is a beautiful example of an extremely simple Rule that leads to a highly evolved structure from absolutely meager beginnings – one live cell. If the pattern in Fig. 5.14 looks familiar, it is because it is also associated with what is known as Pascal's triangle, for those who encountered it in high school. The output is governed again by a simple Rule involving the state of its two neighbors (Fig. 5.13):

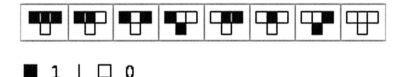

Figure 5.13: Rule 18 with only two configurations that result in live output. Yet hidden away is a remarkable complexity that unfolds with the time evolution of the cellular automaton. (Image courtesy of Wolfram Alpha.)

Just two configurations lead to a live (black) cell. Every other configuration of a central cell and its two neighbors yields a dead (white) output. Yet the time-step evolution of Rule 18 starting from a single live (black) cell is the highly geometric structure of Fig. 5.14[8]. Actually, several different one-dimensional Rules generate this same configuration, if one starts with one black cell and all the others white.

---

[8] Often, Rule 90 is taken as an illustration of this particular output from a single live cell, but the time-step evolution is exactly the same as for Rule 18, although the rules are completely different. A few other rules also generate the same output from a single live cell.

Note also that if we list the output of Rule 18 using 0 (white) and 1 (black), we obtain: 00010010. This is just the number 18 in base 2. Clever. This applies similarly with all the 256 one-dimensional rules. The systematic numbering of these rules derives from this binary representation of the output of the eight possible neighborhoods[9].

We can readily verify that the highly evolved structure in Fig. 5.14 follows from the very simple Rule 18. For example, the top live (black) cell has two dead (white) neighbors and so at the second generation it will be dead (white). But if you look at the dead cell to the left of the live one, it has a single live neighbor and will thus be alive at the second generation. Likewise for the right-hand neighbor, which will also become alive.

So it goes on. Generation after generation, the same pattern will continue to grow and grow. Luckily, the computer can think so much faster than us and will produce this pattern in an instant for as many time-steps as we choose.

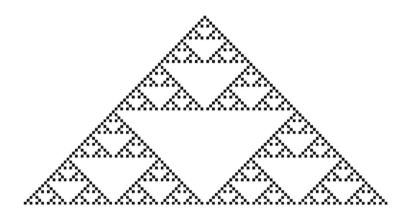

Figure 5.14: The time-step evolution of Rule 18, starting with only one live (black) cell. The CA evolves indefinitely downward in this pattern. (Image courtesy of Wolfram Alpha.)

Rule 18, like many of its counterparts, has a certain complexity about it. Moreover, the smaller triangles with black-celled edges contain triangles within themselves that mimic the features of the main triangle. This is a feature known as *self-similarity*, whereby a part of the whole looks like the whole itself. This cellular automaton duplicates the features of the geometric structure known as a *Sierpinski triangle*.

This famous triangle can be obtained directly by taking an initial equilateral triangle and marking the mid-point of each side and then joining the three marks with three straight lines. This will create three new smaller triangles – one at the

---

[9] It should be mentioned that the neighborhoods are always displayed in exactly the same sequence from left to right.

top and two at the bottom. Taking each smaller triangle in turn, we mark the midpoint of each of the sides and join the marks with straight lines, and so on. By repeating this procedure, you can see that you will arrive at the same structure as in Fig. 5.14, only created in lines instead of small boxes.

On the other hand, if we ran Rule 18 for enough iterations, the boxes would be so small as to be indistinguishable from straight lines. Later in the chapter, we will find that the 'dimension' of the Sierpinski triangle lies between 1 and 2.

It would seem that we always obtain some structured output via our 256 CA rules. This is actually not the case, and we can easily achieve chaotic output in our cellular automaton world. An example is again provided by Rule 18, as in Fig. 5.15.

In this instance, instead of having only one black cell to initiate the automaton, a random configuration of black and white cells has been chosen. The resulting output is completely chaotic. One striking feature of a chaotic system is that if we vary the initial conditions in the slightest, in this case the state of just a single cell, say from 0 to 1 (white to black), then the entire output will be completely altered. This exquisite sensitivity to initial conditions is one key feature of systems that yield chaotic output.

The notions of regularity and chaos will re-appear in the later section regarding Dynamical Systems.

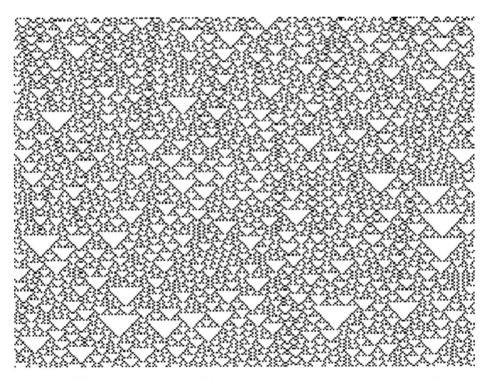

Figure 5.15: Rule 18 again, but this time starting with a random configuration of black and white cells. The result is completely chaotic and although triangles do appear, they appear in a random fashion. The slightest variation in initial conditions will produce different output. (Image courtesy of Wolfram Alpha.)

Could the entire Universe be one gigantic cellular automaton, with the laws of Physics playing out according to the tick of some celestial clock? This may not be as crazy as it may seem, as the quote by Richard Feynman at the beginning of the chapter indicates.

THE WHOLE IS GREATER THAN SUM OF ITS PARTS

Now that we have seen how very complex structures can arise from very simple premises, let us now look to Nature to see what can be gleaned about the organization of complex structures.

We humans are all composed of atoms, which make up molecules, which in turn form proteins, which are the building blocks of us all. At some stage, this sequential construction becomes conscious of itself and we become you and me. But how does a collection of molecules become aware of itself? This is the question of consciousness[10], and there are few answers, although the question has exercised many great minds for centuries. Although there are a vast number of theories regarding consciousness, one is that it arises as an *emergent* phenomenon due to the complexity of the neural connections in our brain.

This view would classify the brain as a *complex adaptive system*; that is, the brain has properties such as consciousness that can emerge and which are not fully explained by understanding the individual connections and chemistry regarding every neuron. In essence, the whole is greater than the sum of its parts.

This circle of ideas regarding the emergence of properties from complex adaptive systems is part of the new Science of Complexity. A few of these systems that also have the ability to self-organize will be presented in what follows. A very readable account can be found in the book by Roger Lewin listed in the Bibliography.

For comparison purposes, let us first consider the Gothic-styled cathedral that was begun in the city of Cologne, Germany in 1248 and was finally completed 632 years later in 1880, adhering to the original medieval plans and drawings (Fig. 5.16 L). This could have only come about because there was a *master plan* handed down to those who continued the work of previous generations. The finished Cologne cathedral is truly a wonder to behold, especially standing inside and looking up and up and up.

---

[10] *Consciousness* is notoriously difficult to define and no one really knows what it is. Here we simply consider consciousness as an ability to sense and respond to both internal and external environments. Various attempts have been made to formulate a mathematical framework for consciousness in recent years; see the article by J. Kleiner, *Mathematical Models of Consciousness*, https://arxiv.org/abs/1907.03223v2/ April 2020.

Figure 5.16: (L) The majestic Cologne Cathedral that was begun in the year 1248 and finished in 1880. (R) Bacteria colony with a high degree of self-organizing complexity. (Images courtesy of (L) Thomas Wolf, www.foto-tw.de, CC BY-SA 3.0 de, (R) Eshel Ben-Jacob.)

BEES AND TERMITES

When it comes to insect populations such as honey bees, ants and termites, however, the situation is both similar and uniquely different. Analogous to the case of the Cologne cathedral, a beehive has many constituent parts and includes a social structure of queen, drones and workers.

The queen's sole function is to produce eggs. The worker bees, which are all females, go out and forage for pollen and nectar to bring back to the hive, as well as doing all the housekeeping such as feeding the queen and larvae. Their lifespan is only about six weeks. The drones mate with the queen in mid-air, nothing more, and usually perish after mating. The queen stores up male sperm that is used selectively to fertilize eggs, or not. Therefore, the sex of the offspring is controlled by the queen, with an unfertilized egg giving rise to a male drone (*parthenogenesis*) and a fertilized egg producing a female worker.

The daily activity of a hive has more action than an episode of *Game of Thrones*. New queens are created by feeding larvae solely on royal jelly and they will fight to the death until one, and only one, remains triumphant in the hive. Elderly, no longer adequate queens are surrounded by worker bees, causing the queen's body temperature to rise until she dies.

The great distinction with the Cologne Cathedral is that there is *no master plan* for this hive of activity. Every constituent part, including the perfectly structured hexagonal honeycomb cells, is performing a specific task that leads to the complex self-organizing system of the beehive[11].

---

[11] The executive wing of the New Zealand Parliament is built in the shape of a beehive (and called the Beehive), no doubt to give the impression of the fevered activity of honey bees.

Similarly, a termite mound can have over a million inhabitants and is a very sophisticated complex structure, sometimes reaching several meters in height. But where is the master plan for all of this?

Living complex systems abound in Nature, yet there is no master plan for any of them and they are considered examples of emergent behavior. As mentioned above, it has been suggested that human consciousness is an emergent phenomenon, as perhaps is life itself – the product of the laws of Physics, Chemistry, Biology, and random events.

… AND ANTS

Ants are very interesting creatures, who build colonies with graveyards for their dead, and always turn up at the nearest picnic even without an invitation. But equally interesting is a particular digital variety devised by scientist Christopher Langton in 1986[12]. These are virtual ants, who live on an array of cells that are initially all white, and can only move in a *N, S, E, W* direction from cell to cell. An ant's behavior is defined by two very simple Rules:

   i. *If an ant lands on a white cell, it changes it to black, then turns 90° to the right and moves one cell forward.*
   ii. *If the ant lands on a black cell, it changes it to white, then turns 90° to the left and moves one cell forward.*

The first eight steps of Langton's ant are depicted in Fig. 5.17.

Making the grid space initially all white and sufficiently large so that the ant is free to move about unencumbered by boundaries, it wanders around aimlessly for thousands of steps in all directions, as one would expect. The ant re-colors cells it has previously visited and strikes out into new territory in a seemingly random fashion.

What is completely unexpected is what happens after some 10,000 steps (digital ants never tire).

The ant falls into a sequence of period 104 in its behavior along a diagonal, thus creating a 'highway' that continues indefinitely. This highway always seems to arise, even if the initial configuration of cells contains a finite number of black cells (Fig. 5.18)[13]. Once the highway begins, the system becomes completely ordered, much like the periodic (cyclic) behavior exhibited by certain configurations in the Game of Life model. However, it should be mentioned that Langton's ant is not a proper cellular automaton, as it does not update all cells simultaneously.

---

[12] These ants were independently discovered by Greg Turk who called them 'termites.'

[13] An excellent implementation of the Langton ant can be found at: https://kartoweb.itc.nl/kobben/D3tests/LangstonsAnt/ where it is possible to see the highway forming after a few minutes of chaotic wanderings by the ant.

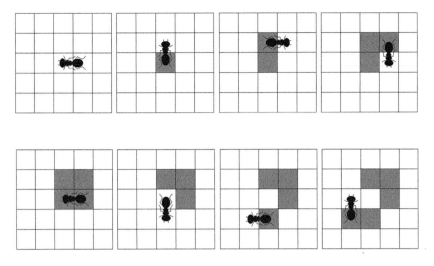

Figure 5.17: This is the Langton ant taking its first eight steps according to the two simple Rules governing its behavior. By the fifth step, the ant has run into a cell it has previously encountered and must turn the black cell back to white and turn left, then off it goes on its aimless journey. (Image courtesy of John Wiley & Sons, reprinted with permission.)

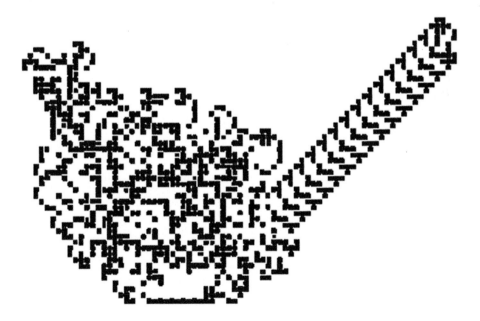

Figure 5.18: The periodic behavior that forms the Langton highway starts at some point after 10,000 steps of aimless wondering, but now the ant will continue constructing the highway indefinitely. Order from chaos.

Somehow, the orderly periodic behavior of the Langton ant is an *emergent* feature of the two simple Rules of ant behavior. Nothing in the Rules suggests that any orderly behavior would ever result, and if the ant was run for only a few thousand steps, it would exhibit exactly the random behavior one would expect of it. Yet out of complete chaos (not defined here in any technical sense), we eventually evolve a completely ordered structure. In the words of theoretical biologist Stuart Kauffman, "Order for free."

What happens when we consider two Langton ants on the same grid space with the same two Rules applying to each? All we need to add is a right-of-way Rule for when the two ants might end up on the same cell.

> iii. *Labelling the ants, A and B, A always has the right-of-way and goes first, followed by B.*

Of course, what plays out over time depends on the relative positions and directions of the two ants with respect to one another. In some scenarios the ants will wander around for a while, each creating its own separate chaos, but will subsequently undo each other's work completely and then return to the position of the other ant but facing in the opposite direction. Following them further, the two ants will carry on as the first time, with each ant undoing the other's work and finishing off at exactly the same position as they started, thus exhibiting a periodic *2-cycle* behavior. Even well-established highways can be undone by the other ant in this fashion.

Using a particular starting arrangement, it is possible to get the two ants to construct two separate Langton highways going off in different directions, after much intermingling with one another. It is almost as if the two ants discuss doing this during their brief encounters, and highway construction occurs much sooner than in the single ant case. Similarly, with just the right initial set-up, former student Malcolm Walsh found that the two ants can actually 'cooperate' on the construction of a single period 102 highway (Fig. 5.19) that takes on a different form to the Langton one.

It has also been observed, in the case of three ants with a suitable right-of-way Rule, that after a spell of intermingling between the three, two of the ants will have gone off together to construct a Walsh highway of period 102, while the third ant will go off to form its own separate Langton highway after about 3,000 steps. This sort of behavior is simply stunning, and it is difficult to believe that the ants are not exhibiting some sort of intelligence. But of course, they are merely following very simple Rules and nothing more. Nothing in the Rules could predict the subsequent construction of highways.

All of this beautiful emergent behavior and complexity from the simplest of Rules of ant behavior.

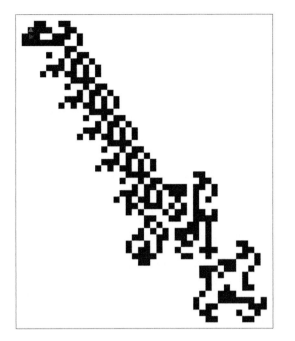

Figure 5.19: The cooperative Walsh highway of period 102, formed by two cooperating ants, that appears after a few hundred steps. The current activity is at the top left. Grid lines have been suppressed. (Image courtesy of John Wiley & Sons, reprinted with permission.).

BACTERIA COUNT

Having considered bees, termites and ants (real and digital), what about bacteria? Their self-organizing colonies can also exhibit a high degree of complexity and symmetry with no overall plan for construction (see Fig. 5.16 R). Do they exhibit any other signs of what we might loosely describe as intelligence? It is certainly true that most parents are quite thrilled when their young child is able to count to 10, which they consider an intellectual milestone. Well actually, bacteria can do something analogous to counting, making them pretty clever as well.

It turns out that bacteria can communicate with other members of their species, and their language is mediated by signaling molecules that they release into their local environment. The higher the concentration of the molecules detected, the larger the local population of bacteria. This ability of bacteria to sense the number of other bacteria in their local environment is a process known as *quorum sensing*.

When the number of bacteria reaches a critical threshold, it allows the colony members to coordinate various physiological responses when placed under environmental stresses, such as a lack of nutrients either through scarcity or competition with other species. This insures the greater survivability of the colony as well as the individual. All for one, one for all – even bacteria understand this principle. "Presumably, this process bestows upon bacteria some of the qualities of higher organisms."[14]

So, bacteria can not only organize themselves into highly complex structures, but they can count, and so form a part of our Mathematical Universe (Fig. 5.20).

It is also important for bacteria to distinguish their own kind from other bacterial species in their environment, and they can do this as well. This is done by the bacteria sending out a cocktail of quorum sensing chemicals. Deciphering the blend of chemicals detected by each bacterium tells them the number and types of species in their local environment. This "allows bacteria to carry out appropriate quorum-sensing-controlled behaviors both in mono- and multi-species communities."[15]

This is now a rather sophisticated form of calculation.

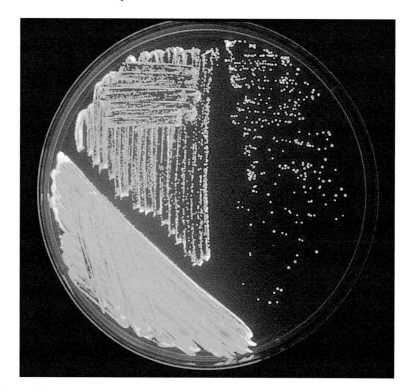

Figure 5.20: The bioluminescent marine bacteria *Viberio harveyi* that uses quorum sensing to decide whether or not to switch on their light. Only when a sufficient concentration of bacteria is achieved do the bacteria initiate a light-generating chemical reaction. Experiments with a related bioluminescent bacterium *V. fischeri*, begun in 1970, are what led to the discovery of quorum sensing. (Image courtesy of Bassler Lab, Princeton University.)

---

[14] M.B. Miller, B.L. Bassler, Quorum Sensing in Bacteria, *Ann. Rev. of Microbiol.*, 55, 165–199, 2001.

[15] B. Bassler: https://explorebiology.org/learn-overview/cell-biology/quorum-sensing:-how-bacteria-communicate

On the other hand, many disease-causing bacteria rely on quorum sensing to do their nasty little work better. Disrupting bacterial quorum sensing via synthetic compounds is one approach to diminishing their pathogenic effectiveness.

We note that as well as bacteria, and even some viruses too, the complex society of honey bees also includes quorum sensing to locate the best new site for a swarm to form a nest. The decision is based on reports (via a 'waggle dance') from numerous (female) worker bee scouts on a multitude of site options. Only when a quorum of 10–15 or more scout bees has been reached, regarding the high suitability of one particular site, is that site chosen for the swarm's new home.

Not be left out, certain types of ants also use quorum sensing based on scout reports to determine the most ideal site for a new nest. Both ants and honey bees are solving a certain type of mathematical problem, one of *optimization*, or how to find an optimal new nesting site among all those available. Each potential site will have a number of properties that affect its suitability as a future home and it is through quorum sensing that the problem of selection is solved.

The reader might ponder just how a large population of human beings might try to solve the same optimization problem of finding the most ideal site for a new home. Perhaps our prehistoric ancestors were faced with just such a problem. Once mankind has to leave Earth, perhaps we will be faced with it again.

A HIVE OF MATHEMATICS: FIBONACCI

Let us consider the honey bees again for a moment, as there is some further interesting Mathematics going on amongst all the buzz. As mentioned, a male bee comes from an unfertilized egg, and thus has only one parent, the queen. However, a female bee comes from a fertilized egg and thus has two parents, the sperm-donating male and the female queen. This leads to a very interesting family tree for a male drone bee.

Let us use 'M' for male bee, 'F' for female bee, and consider the family tree of a single male. Starting with a single M, it comes from an unfertilized egg, so has one F mother. The mother has M+F parents, which are two grandparents of the original M. The M grandparent has one F parent (great grandparent of original M), and the F grandparent has M+F parents (great grandparents), making three great grandparents. To make this easier to consider, let us just make this into a proper family tree, as shown in Fig. 5.21.

If we now total up the number of bees in each generation, we have

$$1, 1, 2, 3, 5, 8, 13, 21, 34, 55, \ldots$$

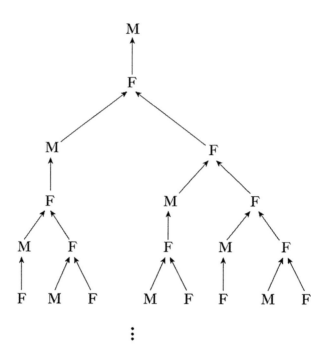

Figure 5.21: The family tree of a male (M) drone bee that comes from an unfertilized egg and so has one female (F) parent. On the other hand, each female in the lineage comes from a fertilized egg and so has both M and F parents. The number of descendants at each generation turn out to be the Fibonacci numbers. (Illustration courtesy of Katy Metcalf.)

where the author has taken the liberty to fill in a few more numbers, as the sequence has a specific pattern. Indeed, the sequence is known as the *Fibonacci sequence*, after the Italian mathematician Fibonacci (*ca.*1175 − *ca.*1250), although the sequence was also known to Indian scholars prior to this. Fibonacci was interested in a hypothetical breeding situation from a pair of rabbits, but the bees are a more realistic example.

The Fibonacci numbers have the simple property that each number (starting with 2) is the sum of the two previous numbers. So, if F_n is the *n*th Fibonacci number, and just setting $F_0 = 0$ for the sake of completeness, then we have the relation

$$F_n = F_{n-1} + F_{n-2}, \qquad (16)$$

for $n = 2, 3, 4, 5, \ldots$

Remarkably, these numbers have some special properties. As was noted by astronomer Johannes Kepler (see Chapter 6), the successive quotients of the Fibonacci numbers, such as: $\frac{3}{2}, \frac{8}{5}, \frac{13}{8}, \frac{21}{13}, ...$, approach the value of the *golden ratio (golden mean)* φ (phi); that is,

$$\frac{F_{n+1}}{F_n} \to \varphi,$$

as $n \to \infty$.

The golden ratio, $\varphi = \frac{1+\sqrt{5}}{2} = 1.61803. ..$, dates back to the ancient Greeks, and arises from the consideration of a line segment divided up into two segments of length a and b, in such a way that the ratio of a to b is the same as that of, $a + b$ to a, or symbolically,

$$\frac{a}{b} = \frac{a+b}{a}.$$

In this instance, the common ratio value is called the golden ratio, φ. In order to find this value, we set $b = 1$, which allows us to work out the value of a from the preceding equation. This yields

$$a^2 = a + 1,$$

and solving this by the quadratic formula, we get

$$a = \varphi = \frac{1+\sqrt{5}}{2}.$$

The reciprocal of the golden ratio has the value,

$$\frac{1}{\varphi} = \frac{-1+\sqrt{5}}{2} = 0.61803...$$

Now here is a gift from the gods concerning φ:

$$\varphi = 1 + \cfrac{1}{1 + \cfrac{1}{1 + \cfrac{1}{1 + \cfrac{1}{1 + \cfrac{1}{1 + \cfrac{1}{1 + \ddots}}}}}}$$

This is known as a 'continued fraction', which carries on *ad infinitum*. This remarkable expression holds because if you examine its structure carefully you will find that it can be written in the form:

$$x = 1 + \frac{1}{x},$$

which is just the relationship satisfied by φ. Is that not beautiful?

The esteemed φ value has appeared in such human applications as architecture (Le Corbusier – Charles-Édouard Jeanneret), art (Salvador Dalí), and music (Mozart), and even the stock market, where φ and its reciprocal, $1/\varphi$, feature in the study of *Fibonacci retracements*.

Moreover, it occurs naturally in the physical world, in the manner in which plants develop their leaves (known as *phyllotaxis*), and has been found in an investigation into the state of linked magnetic atoms of cobalt niobate, put into a 'quantum critical' state[16]. According to one of the authors, "Such discoveries are leading physicists to speculate that the quantum, atomic scale world may have its own underlying order." This is particularly interesting, as behavior in the quantum world has little correspondence with the behavior of physical systems in our macro world.

According to physicist Vladimir Pletser, "Fibonacci numbers ... and the golden ratio can be found in nearly all domains of science, appearing when self-organization processes are at play and/or expressing minimum energy configurations[17]."

DYNAMICAL SYSTEMS

The *recurrence relation* of the Fibonacci sequence of numbers, as given in eq. (16), is an example of a *Dynamical System,* and as we saw, the number of ancestors of a male drone bee followed the formula. This is the kind of system we consider now, namely a population of some sort, whose size at one generation depends on its size in the previous generation.

Therefore, let us represent the population size at year zero (its initial size) by x_0, at year one by x_1, year two by x_2, and so on. Then if, say, the population doubles from one year to the next, we can express this as

$$x_{n+1} = 2x_n, \qquad n = 0, 1, 2, 3, \ldots$$

---

[16] R. Coldea, *et al.*, Quantum criticality in an Ising chain: experimental evidence for emergent E8 symmetry, *Science* 327, 177, 2010.

[17] Chinese Academy of Sciences, Fibonacci Numbers and the Golden Ratio in Biology, Physics, Astrophysics, Chemistry and Technology: A Non-Exhaustive Review, PDF.

In other words, whatever the population is in year n, the population will be double the next year, year $n + 1$. Equations of this nature are found in Biology where things grow, as well as many other branches of Science, including Meteorology where we will encounter the 'butterfly effect', as in the concept of chaos. Unfortunately, the equations are not that simple, unless of course a population really does double from year to year.

One equation that encompasses all the features of interest is the so-called *logistic equation (map)*, a discrete population growth model based on an equation developed by Belgian mathematician Pierre Verhulst (1804–1845):

$$x_{n+1} = ax_n\left(1 - x_n\right), \quad n = 0, 1, 2, 3, \ldots, \tag{17}$$

where a is just a constant parameter, with $0 < a \leq 4$, and the numbers x_n represent population values in the year n.

The behavior of this system from one year to the next is dependent on the value chosen for the parameter a, and it turns out that this behavior has very interesting features. This model received prominence with the article "Simple mathematical models with very complicated dynamics" by mathematical biologist Robert M. May (now Baron May of Oxford), published in the journal *Nature* in 1976[18].

Any parameter value $a < 1$, has the iterates quickly diminishing in value to zero, representing *extinction*. So, this range of the parameter a is rather boring. For values of the parameter a with $1 < a < 3$, the iterations converge to a single fixed value, which is given by $(a - 1)/a$.

For example, let us start with the parameter value $a = 2.75$, and some initial population value, say, $x_0 = 0.001$. The initial value does not affect the long-term behavior of the system, and is only of significance when the system becomes chaotic (discussed below).

Of course, *real* populations have whole number values, but this is just a model and our population values are always going to be less than 1 for input and output, which also serves to keep the numbers manageable. The calculations are best done on a computer and the first result can be seen in Fig. 5.22[19]. The values x_n are approaching (converging to), $1.75/2.75 = 0.63636\ldots$, which is called an *attractor* of the system, and the value 0 is a *repeller*.

In general, the sequence of x_n values generated from a given initial point is called an *orbit of* x_0, in analogy with planetary orbits which have fixed positions at each instant of time.

---

[18] There were prior discrete models such as the *Ricker* model that is applicable to fisheries: $x_{n+1} = e^{a(1-x_n)}$, which exhibits analogous behavior.

[19] The required computer code is very short and simple and it is left to the reader if they wish to pursue it further.

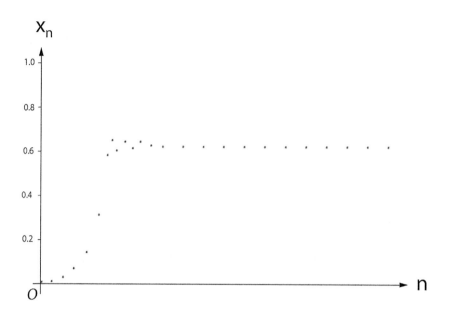

Figure 5.22: Iterations of the Dynamical System $x_n = a\, x_n(\, 1 - x_n)$, *with a* = 2.75 and initial value $x_0 = 0.001$. The values x_n are approaching the attractor 0.63636… (Illustration courtesy of Katy Metcalf.)

More interesting behavior is displayed as we increase the value of the parameter a, for example taking $a = 3.2$. In this instance, the values of the Dynamical System oscillate between two values 0.5130 and 0.7994 (to four decimal places), and are said to converge to an attractive *2-cycle*. The system is said to be *periodic* with *period two*. We witnessed just such periodic behavior in the Game of Life, seen in Fig. 5.5 (top line).

Upping the ante somewhat, at $a = 3.5$ we obtain a 4-cycle, or period four attractor, that oscillates between the values 0.5009, 0.8750, 0.3820, 0.8269 (to four decimal places), where the values of x_n appear and repeat in that order (Fig. 5.23). There are also period eight attractors and so forth, in a region where increasing the value of the parameter a causes the system to go through a phase of period doubling, 2, 4, 8, 16… We will encounter just such period doubling again in our subsequent discussion of the Mandelbrot set.

Lest one think that our Dynamical System only produces even cycles, with a bit of delicate probing one can find a period three attractor at $a = 3.83$ (Fig. 5.24), and of course further probing yields other odd cycles as well. The author once offered a financial reward to any students in a class who could find odd period attractors higher than a 3-cycle. This proved to be a costly exercise.

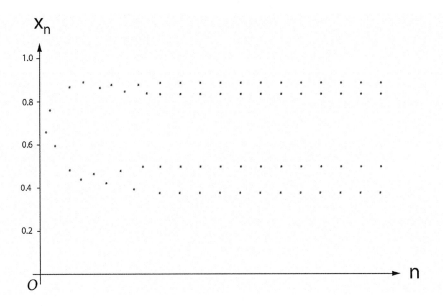

Figure 5.23: A 4-cycle with $a = 3.5$ that alternates between the four values 0.5009, 0.8750, 0.3820, 0.8269 (four decimal places), respectively. (Illustration courtesy of Katy Metcalf.)

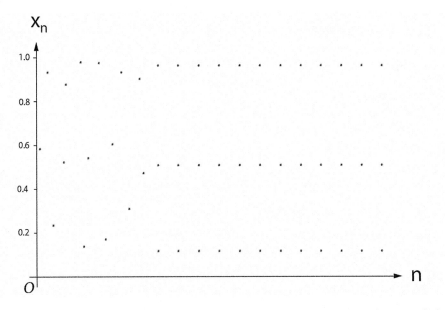

Figure 5.24: After a brief spell of unsettled activity, the system settles into a period three attractor found at the value $a = 3.83$, giving values 0.1561, 0.5047, 0.9574 (four decimal places). (Illustration courtesy of Katy Metcalf.)

At the end of our parameter interval, $a = 4$, and at some other values as well, we have what is known as *chaos*, and one feature is the completely random nature of the output (Fig. 5.25). Another very important feature of a chaotic system is that the output is exquisitely sensitive to the initial conditions as represented by the initial value of x_0. This sensitivity to initial conditions was also mentioned in the context of the chaotic output of Rule 18 (Fig. 5.15) and here we find it once again.

To see this, let us take the initial value $x_0 = 0.987654321$. The value after 100 iterations is then:

$$x_{100} = 0.541650585287708.$$

Now, taking the slightest perturbation of the initial x_0 value, say $x_0 = 0.987654320$, a change of one-billionth, we obtain:

$$x_{100} = 0.979506807437642,$$

which is not even close to the previous value for x_{100}.

Indeed, the respective orbit values of x_n are very different for many prior iterations as well. Only the very first few iterations of their respective orbits will be close in value for the two choices of x_0.

Observations such as these were made by the research meteorologist Edward Lorenz in the early 1960s, when he was studying a system of equations to make weather predictions. One day, he put values into his equations that were rounded off figures from a previous simulation. The subsequent weather predictions in due course diverged widely from the previous forecast. This extreme sensitivity of a chaotic system to initial conditions became fancifully known as the *butterfly effect*, in which, so to speak, the flapping of a butterfly's wings in Brazil could set off a tornado in Texas.

As a consequence, short term predictions are possible, as say for the weather (which is a chaotic system) over the next few days, but long-term predictions become increasingly out of the question.

Another interesting feature of the chaotic behavior that ensues with certain values of the parameter a, is that the system is completely deterministic. The same feature was found with our time-step iterations of certain Cellular Automata (Fig. 5.15). Here, we have a very simple formula for how to compute every value from the preceding one. Under certain choices of the given parameter, we obtain very regular periodic behavior. With some other choices, the behavior becomes completely random and unpredictable in the long term, even though the next value x_{n+1} of the system is rigorously defined by eq. (17).

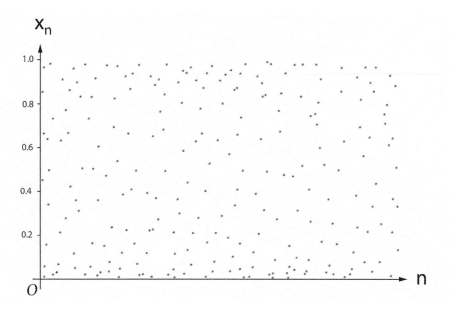

Figure 5.25: Putting $a = 4$ into our Dynamical System: $x_{n+1} = ax_n(1 - x_n)$, $n = 0, 1, 2, 3$, ..., results in chaos. Here we have taken more iterations of the system than in the preceding examples. (Illustration courtesy of Katy Metcalf.)

MESSRS. FATOU, JULIA, AND MANDELBROT

We took real numbers in our logistic map, but of course we could also consider Dynamical Systems where the components are complex numbers. In this case, we again find remarkable complexity hidden within the simplest of expressions.

Let us consider a very elementary Dynamical System[20]:

$$z_{n+1} = z_n^2 + c, \quad n = 0, 1, 2, 3,... \tag{18}$$

where c is a constant *complex* number of our choosing. Since squaring a complex number also squares its modulus[21], we have to be careful in our choice of initial starting value z_0, or else the iterates of z_n will simply run off to infinity, which is not very interesting.

Thus, we will confine our attention to those (seed) values of z_0 which lead to values that remain bounded; that is, they do not run off to infinity, no matter how high a value of n we take. It turns out that we only have to look at iterates that remain within a circle of radius = 2, for if it goes outside this circle then there is

---

[20] Although this Dynamical System appears remarkably simple, its dynamics are anything but simple, and we consider it here for historical reasons.

[21] Writing $z = re^{i\theta}$, then, $z^2 = r^2 e^{i2\theta}$, so that squaring a complex number z, squares the resulting modulus r, which is the distance from the origin. Note that the argument θ is doubled.

no stopping it, and z_n values are off to infinity as $n \to \infty$. Again, a computer is useful for the purposes of doing the arithmetic of the iterations, once we have chosen the constant value c and a starting point z_0.

Collecting up the set of all starting points z_0 whose iterates *do not* go off to infinity is known as the *filled-in Julia set*. Its boundary is the *Julia set,* although sometimes the two notions are conflated. Any point outside this figure will ultimately diverge to infinity under iteration of our quadratic eq. (18), and this external region is referred to as the *Fatou set* (Fig. 5.26)

Both Julia and Fatou sets were considered by French mathematicians Pierre Fatou (1878–1929) and Gaston Julia (1893–1978). The latter lost his nose due to an injury incurred during World War I, and subsequently wore a leather patch over it. One of his students in Paris was Benoit Mandelbrot. Although Julia himself did not have a computer to display his magnificently complex sets, if he had, he would have seen that his sets are beautiful examples of *fractals,* as will be discussed in the next section.

Figure 5.26: (L) The filled-in Julia set (black region) for the Dynamical System : $z_{n+1} = z_n^2 + c$, for $c = -0.13 + 0.68i$, a value chosen by the author. The exquisitely intricate boundary is the Julia set and the complement represents the Fatou set whose iterations go off to infinity. The more the boundary is magnified, the more it continues to unfold its intricate self-similarity. The red/orange coloring indicates longer escape times to infinity than points in the blue region (see also Fig. 5.27). (R) A Julia set again, for $z_{n+1} = z_n^2 + c$, for $c = -0.18 + 0.66i$, but in this instance the set has no interior. (Image source: (L) Created with James Paterson's Julia set viewer at http://thejamespaterson.com/scripts/julia/, (R) Courtesy of Chris King.)

Of course, we are not restricted to a simple quadratic for our Dynamical System. We can take more general quadratics, cubics, and trigonometry functions like *sine* and *cosine,* as well as the exponential function,

$$z_{n+1} = e^{z_n} + c, \quad n = 0, 1, 2, 3, \dots$$

and complex constant c. The results are even more impressive (Fig. 5.27).

Figure 5.27: The filled-in Julia set (in black) for $z_{n+1} = e^{z_n} + c$, and $n = 0, 1, 2, 3, \ldots$, with $c = 0.2 + 0.3i$. Colors indicate the different 'escape times' for the points outside the Julia set and represent the number of iterations required to become larger (in absolute value) than a particular set value, on their way to becoming infinite. The colors toward the center of the spirals indicate that more iterations of the function are needed for that particular seed value of z_0. The scale is: $-2 \leq x \leq 2$; $-2i \leq iy \leq 2i$. (Image created with James Paterson's Julia set viewer at http://thejamespaterson.com/scripts/julia.)

Many have heard of and seen the set that comes next, namely the Mandelbrot set (Fig. 5.28), another set of infinitely exquisite complexity, and one of the most interesting sets ever discovered[22].

The Mandelbrot set is a visual representation of all the values of the complex constant c that produce bounded iterations of our same eq. (18), but in this instance, starting with the initial seed value $z_0 = 0$ every time. Again, once the modulus of an iteration exceeds the value of 2, it is headed to infinity and there is no need to compute further values. Knowing this saves a lot of computer time. Each fixed value of c in the Mandelbrot set has a corresponding Julia set in the sense described above.

---

[22] Precisely who should get credit for the discovery of the Mandelbrot set is an issue as complex as the set itself. See the article by John Horgan: Who Discovered the Mandelbrot Set, *Sci. Amer.*, March 2019. The set *does* appear explicitly in the work of mathematicians J. Peter Matelski and Robert Brooks prior to publication by Mandelbrot, but as usual, it is complicated.

Benoit Mandelbrot (1924–2010) was a mathematical scientist and is considered the "father of fractals". On account of his work on fractals, the Mandelbrot set was named after him. The quadratic equation that Mandelbrot had been working on in conjunction with his eponymous set was none other than our former logistic map:

$$z_{n+1} = a\, z_n \left(1 - z_n\right), \qquad n = 0, 1, 2, 3, \ldots$$

in terms of complex numbers a and z_n. The form of eq. (18) is even simpler, yet the complexity it generates is astounding. As it turns out, eq. (18) and the preceding complex logistic equation are algebraically related.

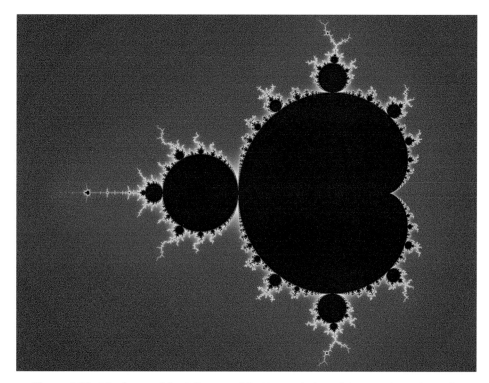

Figure 5.28: The famous Mandelbrot set (black), consisting of the complex values of c for which the iterations of $z_{n+1} = z_n^2 + c$, remain bounded, when starting with the initial value $z_0 = 0$. The cusp at the right is on the real axis at $x = 0.25$, and the furthest point to the left lies at $x = -2$. An infinite number of disk-like components attach to the surface of the main cardioid, each of which in turn likewise have disk components, *ad infinitum*. Thin filaments (highlighted in white) link the main structure with outer islands. Magnification of the figure can be carried out indefinitely to reveal an infinite universe of remarkable beauty[23]. (Image courtesy of Wolfgang Beyer, CC BY-SA3.0.)

---

[23] Witness this remarkable beauty and complexity at: https://www.youtube.com/watch?v=i2lDunJZ_pc. Image courtesy of Wolfgang Beyer.

Interestingly, there are regions in the Mandelbrot set for which the iterations z_n not only remain bounded inside the circle of radius 2, but display the values of a periodic attractor as in the logistic map. For example, for the values of c in the circular bulb region (a disk of radius 0.25 centered at the point $x = -1$) immediately to the left of the *main cardioid*, the iterations z_n there converge to a 2-cycle, eventually oscillating between two complex values as $n \to \infty$[23]. Similarly, for the smaller circular bulb to the left of the preceding one, all the values of c lead to the iterations z_n forming a 4-cycle, and so forth for smaller and smaller circular regions, moving to the left in an episode of period doubling that we have already witnessed in the behavior of the logistic equation.

Other primary bulbs around the main cardioid are composed of parameter c values that also yield periodic behavior of the iterates z_n (Fig. 5.29). For instance, for the values of c in the disk-like component at the top of the main cardioid, the

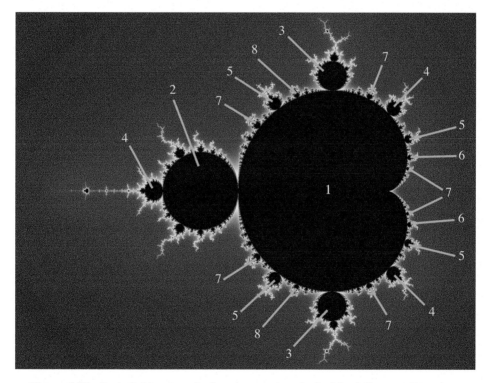

Figure 5.29: Periodicities are to be found everywhere in the Mandelbrot set. For values of c in the main cardioid, their iterations z_n form a 1-cycle converging to a single complex number in each case. Moreover, for values of c in any of the primary bulbs adjoining the main cardioid, we find periodic orbits for each bulb as indicated. Secondary bulbs directly attached to the primary bulbs have periodic orbits that are a multiple of the periodicity of the primary bulb. (Image adapted from Fig. 5.28.)

---

[23] For example, taking $c = -1$ and $z_0 = 0$, then: $z_1 = -1$, $z_2 = 0$, $z_3 = -1$, $z_4 = 0$, ... and so the iterates form a 2-cycle oscillating between the values -1 and 0.

iterates z_n display period three behavior. Furthermore, secondary bulbs attached to primary bulbs exhibit periodicities that are a multiple of the primary bulb's periodicity.

To summarize, the Julia sets arise from eq. (18) by fixing a value of the constant c and finding all the *seed values* z_0 for which the iterates z_n do not wander off to infinity. On the other hand, the Mandelbrot set also arises by considering eq. (18), but fixing the initial value at $z_0 = 0$, and considering all the complex *values of the constant c* for which the iterates z_n (i.e. the orbit of the origin) do not go off to infinity. The two are intimately related due to the fact that each point c in the Mandelbrot set corresponds to a Julia set.

THE FRACTAL UNIVERSE

One feature of most Julia sets is that they possess the characteristic known as *self-similarity*; that is, their appearance is essentially repeated at smaller and smaller scales. This is not, in general, the case with the Mandelbrot set, in that in spite of the infinite degree of detail that is present at ever decreasing scales, what is actually seen depends on the boundary location, although it does exhibit self-similarity in a neighborhood of certain points (*Misiurewicz points*).

Self-similarity can also be found in the cellular automaton Rule 18 (Fig. 5.14), where the triangular regions are reproduced on a smaller and smaller scale. The internet is self-similar in a sense, although one cannot examine the network at an infinitely smaller scale.

Let us see how to describe self-similarity in a more scientific way. We can start with a square sheet of paper and fold it in half, say four times. It will look like Fig. 5.30.

Most people have folded sheets of paper, but perhaps have never thought of the consequences, which are as simple as they are significant.

Now let us forget the piece of paper (which has a thickness), and just consider the grid of 16 squares in the abstract sense of a perfect mathematical object. The length of each side of the smaller squares is 1/4 of the whole square's length (or width). Let us just denote this ratio by $\dfrac{1}{r} = \dfrac{1}{4}$, and call $r = 4$, the *scale factor*[24]. Furthermore, let us note that the dimension of our figure is $D = 2$. So far, so obvious.

But now there is a nice connection between the numbers involved, since we have a 2, 4, and 16, with $4^2 = 16$; that is, $r^D = N$. Interesting, is it not?

---

[24] Sometimes the scale factor is taken to be the reciprocal of what we have defined here, so that the factor $1/r$ ends up in the resulting equations. Either way is fine.

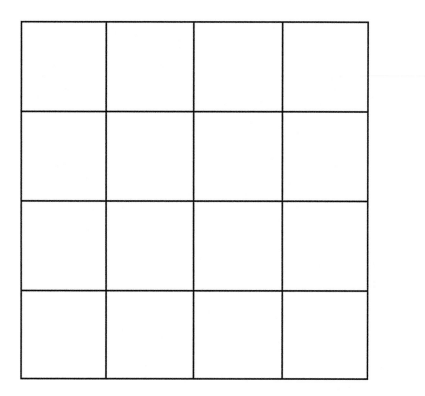

Figure 5.30: By folding a square sheet of paper in half four times, it will look like this where the fold lines have been marked for emphasis.

What if we divided up our square into, say, 64 squares, that is 8 × 8. Then each side of the smaller squares would be 1/8th the side of the big square so that the scale factor is, $r = 8$. Again, the dimension is $D = 2$, and so we have the values, 2, 8, and 64, with, $8^2 = 64$, which is $r^D = N$, yet again. This is no coincidence, and if we continued dividing our square in this fashion, we would always find the same relations between the numbers involved.

Therefore, we have found that

$$r^D = N, \tag{19}$$

where D is the dimension 2 of our space, and N is the number of small representations of our larger figure. This formulation has been right under everyone's noses since people began folding paper; it is simplicity itself. Moreover, the same relationship as in eq. (19) holds if we simply divide up a given line segment into equal length segments in the one-dimensional case, or divide up a cube into smaller ones in the three-dimensional case.

Note that eq. (19) is just a power law of the sort we have already encountered in Chapter 4. One nice feature of power laws for physical systems is that they are *scale-invariant*, in the sense that they hold true over a wide range of values.

Newton's Law of Universal Gravitation (see eq. (28), Chapter 7) is another example of a power law that is applicable to small bodies like humans, as well as large ones like planets and galaxies, with the gravitational attraction only dependent on the mass of the two bodies. We could, in principle, divide up the square indefinitely and yet the same power law as above would apply in every case as N became infinitely larger.

Just as an aside, this paper folding leads to a good party question. Suppose you had a large sheet of paper and folded it repeatedly in half 50 times. How big would the folded paper be: 1 meter, 10 meters, or out to the orbit of Mars?

Well, assuming a sheet of paper is roughly 0.1 millimeters thick, and each time the paper is folded (50 times), the thickness doubles, then the total thickness of the folded paper in mm is:

$$0.1 \times 2^{50} = 1.13 \times 10^{14} \ mm.$$

Since mm tells us little regarding the height, let us convert this to km by dividing our answer by 10^6 mm/km. This gives our paper stack a height of: 1.13×10^8 km, or 113 million km. Mars at its closest is 56 million km from Earth and averages 225 million km. If you timed your folding just right, the stack of paper, and you, could land on Mars. Try this on your friends.

Now back to eq. (19), and our relationship between the scale factor, the number of components, and the dimension. Since the dimension D is in an awkward position as an exponent in the equation, it is more user friendly to take the logarithm (base 10) of both sides and solve for D:

$$D = \frac{\log N}{\log r} . \tag{20}$$

All that from folding a sheet of paper. This newly defined dimension is called the *similarity*, or *fractal dimension*.

As an example, let us compute the fractal dimension of the resulting output from our CA Rule 18 if we ran it indefinitely, so that the rows of cells would look essentially like straight lines. Since each of the three sides of the largest triangle have become divided in half to form three more similar triangles, and so on with the resulting triangles, this means that the scaling factor is: $r = 2$. The resulting number of similar triangles at each stage is: $N = 3$. Therefore, the fractal dimension of the figure in the limit (the Sierpinski triangle) is,

$$D = \frac{\log 3}{\log 2} = 1.585.$$

There we have it, the figure is more than one-dimensional but less than two-dimensional, which is not something we are familiar with. But fractals are extraordinary sets, somewhat other worldly and more at home in the mathematical world, so they must be treated differently. Note that in the case of our square, when $r = 4$ and $N = 16$, then by eq. (20), $D = 2$. In other words, the fractal dimension conforms exactly to the usual dimension in the case of ordinary figures.

There is an interesting real-life application to fractal dimension and that is when trying to measure the length of fractured border of a country. This is well-nigh impossible, as the smaller the measuring device (like a stick) becomes, the longer the boundary becomes. It is not even reasonable in principle to measure around every rock and crevice, since there is no limit on the smallest scale size. So, the very notion of the length of a border is fraught with inherent difficulties.

But a useful concept is to consider the fractal dimension of the coastline instead. We can think of the scale factor r as a magnification factor. In order to do this, you need only to get out a reasonable size map of your country and measure the length of the border by using different length measuring sticks. That is, if the measuring stick length is initially given as say, $r = 1$ unit, then zooming in by subsequent scale factors, say, $2r$, $4r$, $8r$, ..., means that the measuring stick is correspondingly: 1/2 unit, 1/4 unit, 1/8 unit, and so on. In each case, there will be the corresponding (increasing) number, N, of times each length would be needed to approximate the entire border.

The useful aspect of taking the fractal dimension is that we do not have to keep measuring the border length with ever-diminishing length measuring sticks. Note that if we plot our corresponding results of N and r on a graph, where we actually plot the values of $\log N$ on the y-axis vs. $\log r$ on the x-axis, we only need to plot the values of a few points, since the dimension D in eq. (20) is nothing more than the slope of the straight line:

$$y = \log N = D \log r = Dx.$$

In Fig. 5.31, the fractal dimension of the coastline of Great Britain has been computed using four different scale factors. A fractal dimension close to the value 1 indicates a relatively smooth border, and higher values indicated a much more irregular contour.

As we have noticed with the Mandelbrot set and Julia sets, the boundaries are highly fractured, hence they are called *fractals*, and if you know where to look on the boundary of the Mandelbrot set, it will also show self-similarity. The beautiful thing about the Mathematical Universe is that one can look in principle at smaller and smaller scales *ad infinitum*, but in the real world, like the fractured pattern on a leaf's edge, the self-similarity eventually comes to an end (Fig. 5.32 R).

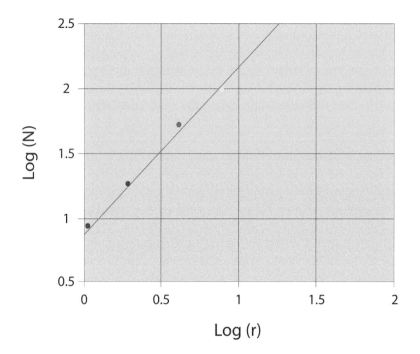

Figure 5.31: The fractal dimension of the coastline of Great Britain is given by the slope of the line, which *is D* = 1.21. For comparison the fractal dimension of the coast-line of Norway is 1.52, and that of South Africa is 1.05. (Image courtesy of Fractal Foundation (fractalfoundation.org).)

Figure 5.32: (L) A self-similar computer-generated fractal that can continue on as long as the computer runs. (R) A tree fern leaf in Rotorua, New Zealand exhibiting self-similarity. (Images courtesy of (L) R. Wicklin, (2014), "Self-similar structures from Kronecker products." The DO Loop blog. URL: https://blogs.sas.com/content/iml/2014/12/17/self-similar-structures-from-kronecker-products.html/ and (R) in public domain.)

In actual fact, computer time comes to an end as well (Fig. 5.32 L). Self-similarity is also allowed the luxury of just being similar at increasingly smaller scales, as the name suggests, not necessarily identical.

Fractal-like scaling has also found its way into recent research concerning the properties of systems of atomic particles as they change over time and space[25].

---

[25] N. Wolchover, The Universal Law That Aims Time's Arrow, *Quanta*, Aug. 1, 2019.

6

Mathematics in Space

... [E]veryone who is seriously involved in the pursuit of science becomes convinced that a spirit is manifest in the laws of the Universe – a spirit vastly superior to that of man, and one in the face of which we with our modest powers must feel humble ...
Albert Einstein

FASTER THAN A SPEEDING BULLET

For many centuries, for those who gave the notion of light any consideration at all, it was thought that light could cover any distance instantaneously, although there were some scholars who thought it had a finite speed. One of the first to have any doubts was Galileo Galilei who, it is said, placed two people with shuttered lanterns some distance apart. When the light from one lantern reached the other party, that person was to open the shutter on their lantern and the time lag would be measured, taking into account reaction times. Unfortunately, any time delay turned out to be impossible to measure, but Galileo was at least on the right track.

Fortunately for Danish Astronomer Ole Rømer, in 1610 Galileo had also discovered the four largest moons of Jupiter: Io, Europa, Ganymede, and Callisto. In fact, anyone with even a small telescope can see these four (Galilean) moons quite easily. If you do observe them from night to night, you will find that they are in different positions because they all orbit around Jupiter.

In the 1670s, Rømer was making a careful study of the eclipse of Io by Jupiter, and found discrepancies in the timing of the eclipses depending on how near or far

J. L. Schiff, *The Mathematical Universe*, Springer Praxis Books, https://doi.org/10.1007/978-3-030-50649-0_6

away the Earth was from Jupiter[1]. He attributed this discrepancy to the difference in the distance that the light had to travel in reaching Earth.

Ever since Rømer, various people have made increasingly more accurate determinations of the velocity of light. The fact that we can even measure something of this velocity is truly astonishing and indicates how far Science has progressed. By the 17th century, the speed of light was known to be finite, and in 1729, English physicist James Bradley calculated that speed to be 301,000 km/sec.

This is exceedingly close to its true value, and for general purposes we take the figure 300,000 km/sec (186,000 mi/sec). This means that a light beam could travel 7 1/2 times around the Earth in a single second. Or that the light from the Sun, which is 150 million km from Earth, takes 8 1/3 minutes to reach us. Technically, of course, the speed of light is measured in a vacuum, but even measured in air the value is nearly the same.

Moreover, the speed of all electromagnetic radiation, such as radio and TV signals, microwave radiation, infrared, X-rays etc., is the same, so it is fortunate that we only have one number to remember. This number, the speed of light, is usually designated by the letter c, as it is constant in value. It pops up in Physics equations in situations that have nothing to do with light, in a similar fashion to how the value π turns up in equations that apparently have nothing to do with circles. The speed of light c is part of the fabric of our Universe. It will play a fundamental role in our discussion of Einstein's Theory of Relativity, which we will come to in the next chapter.

Since the speed of light is so fundamental to our Universe, let us see if we can make some determination of its value ourselves. Given the staggeringly high value of c, this is no mean feat to be sure.

In order to calculate the velocity of light, we first must understand a little bit about electromagnetic waves themselves. These are waves that transmit energy through space − even empty space − unlike sound waves which require a medium in which to propagate. The wavelength is given by the physical distance measured between any two identical points on the wave form, such as the distance between two peaks or two troughs. The wavelength measures one cycle of the wave and is denoted by λ (Fig. 6.1).

Radio waves have very large wavelengths, measured in meters to hundreds of meters. The microwaves in your oven are of the order of a few centimeters in length, while those of visible light range from 380 nm (*nanometers* − billionths of a meter, i.e. 10^{-9} m) at the violet end, to 740 nm at the red end of the light spectrum. Our eyes are (fortunately) attuned to just this part of the broader electromagnetic spectrum. We are unable to see the lower wavelengths of the infrared (*infra* meaning 'below' in Latin) or the higher ones of the ultraviolet (*ultra* in Latin meaning 'beyond'). However, with the appropriate instruments we are able to detect the radiation across all of the electromagnetic spectrum, from radio waves to gamma rays (Fig. 6.2).

---

[1] This distance is constantly changing due to the orbits of the Earth and Jupiter around the Sun.

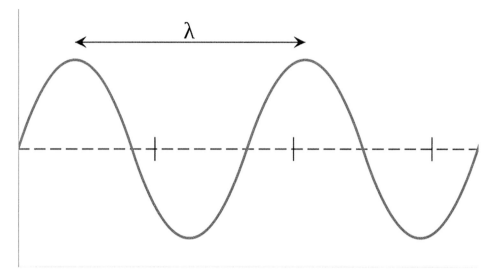

Figure 6.1: The length of a wave is the distance from one crest to another and denoted by the symbol λ. (Image in public domain.)

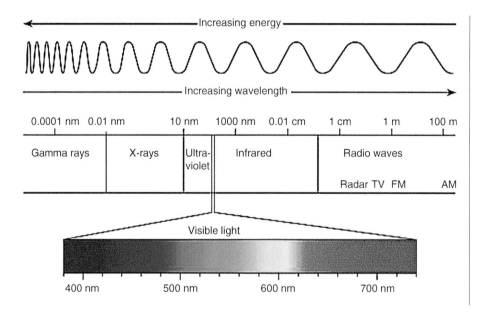

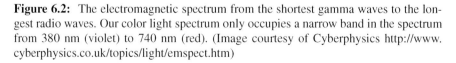

Figure 6.2: The electromagnetic spectrum from the shortest gamma waves to the longest radio waves. Our color light spectrum only occupies a narrow band in the spectrum from 380 nm (violet) to 740 nm (red). (Image courtesy of Cyberphysics http://www.cyberphysics.co.uk/topics/light/emspect.htm)

The energy that is transported by the wave depends on how many waves pass by a reference point every second. So, if you represent the reference point and are counting waves passing you by, then you would be counting how frequently you encountered a complete wave (cycle) each second, that is, you would be determining the wave's *frequency*. The units would be cycles per second.

It is clear from Fig. 6.2 that radio waves have a very low frequency and that X-rays have a very high frequency. If a wave has a frequency of 1 cycle per second, that is, exactly one wave passes a reference point each second, then the frequency is said to be 1 *Hertz*, or 1 Hz, named in honor of Heinrich Hertz, the German physicist who was the first to transmit radio waves.

It turns out that the wavelength, the velocity of the wave and the frequency are intimately related by some simple reasoning. Let us see how by taking the following example.

Just imagine that our wave is now a very long caterpillar that moves up and down like a wave, and that it measures 2 cm from hump to hump (its wavelength) as it moves forward. You put the caterpillar at a starting line (our reference point) and you count, for example, 3 caterpillar cycles each second passing over the line (the frequency). Since 3 cycles pass the line each second, and each cycle measures 2 cm in length, we can say that the caterpillar is travelling 6 cm per second, which is its speed. The upshot from our caterpillar example is that:

$$speed = frequency \times wavelength.$$

Replacing the caterpillar with electromagnetic waves that are travelling at the speed of light, *c*, we have:

$$c = frequency \times wavelength. \tag{21}$$

We have just learned some important Physics simply by conducting a thought experiment about a humble caterpillar, and no animals were harmed in the process.

Equipped with all this knowledge of electromagnetic waves, we are now in a position to measure the speed of light right in our own kitchen. Do try this at home.

The one bit of equipment you will need is a microwave oven. Microwaves, as we know, travel at the speed of light, *c*, and so it will suffice to measure the speed of these waves.

If you look on the back of your microwave oven, you will see the frequency of the waves printed somewhere. Commonly, the frequency is 2,450 MHz or 2.45 GHz, which is the same thing. The 'M' stands for million and the 'G' for billion, and both values represent the frequency of the microwaves: 2,450,000,000 cycles per second. This means that nearly 2 1/2 billion cycles pass by any reference point each second. This is considerably faster than our caterpillar, but the principles are still the same and our eq. (21) now reads:

$$c = (2,450,000,000) \times wavelength.$$

Next, we simply need to find out how big the microwaves in the oven actually are; that is, their wavelength. Then we will multiply this value by 2,450,000,000 and the result will be the speed of light, c.

In order to measure the wavelength, we have to remove the rotating base. Take a dinner plate and cover it completely with either marshmallows or chocolate buttons, depending on taste. Turn on the microwave for several seconds, just enough to melt some of the marshmallows/chocolate buttons ever so slightly. Take the plate out, put it down on a table and mark the position of two adjacent melted points, say with toothpicks. Then measure the distance between these two points in centimeters. If you measure it in inches, just multiply by 2.54 which gives the result in centimeters. Let us say you obtained the value of 6 centimeters between melted points. We are nearly done.

The only thing to understand now is that the microwaves in your oven are what are called *standing waves* and look like those in Fig. 6.3. They are also the same as the harmonics produced from a vibrating string that we saw in Fig. 2.5 in Chapter 2.

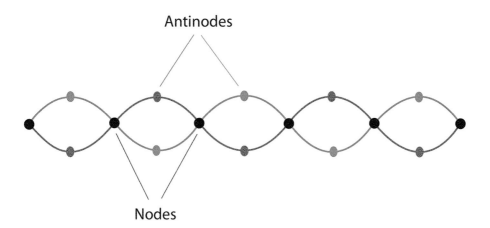

Figure 6.3: A standing wave, illustrating its nodes and antinodes, which is so called as it appears to be stationary, unlike ordinary electromagnetic waves which travel from one point to another. The blue part of the wave moves downward to the red part and back again in a continuously oscillating fashion. At the nodes the wave oscillations cancel one another. (Illustration courtesy of Katy Metcalf.)

These waves are the result of the microwaves repeatedly bouncing off the walls of the oven, with the same frequency and wavelength but reflected in the opposite direction to form this new type of wave that appears to bob up and down and remain stationary. Guitar strings and violin strings also react in the same manner and produce standing waves. The crests are the hottest spots and that is where the toothpick markers will have been placed in the marshmallows/chocolate buttons.

The thing to note about standing waves is that the distance between two adjacent crests (antinodes) represents *half* the wavelength (as in Fig. 6.3), so that our measured 6 cm actually gives a wavelength of 12 cm for the microwaves[2]. Putting this into our equation, it becomes:

$$c = 2,450,000,000 \times 12 = 29,400,000,000\, cm/sec.$$

To convert this answer to meters, we simply divide by 100, so that our speed of light has been determined to be:

$$c = 294,000,000\, m/sec.$$

This is very close to the actual speed of light of $c = 299,792,458$ m/sec, which is taken as the *de facto* velocity to be used in all equations in which it occurs. This universal constant of Nature has literally just popped out of our microwave oven.

As the speed of light is a universal constant, we can use it to explore the rest of the Universe, as will be done later in the text.

DOWN TO EARTH

Through the fields of Astronomy and Cosmology, the Universe itself has become accessible to scientific scrutiny, via Mathematics, right back to its very origins. It turns out that we live in a very Mathematical Universe indeed.

But before leaving planet Earth, let us see how we can measure the Earth's size using some very primitive Mathematics. As we have been witnessing, Science is just rigorous curiosity, so let us examine some early Science, at least in principle, without leaving the comfort of our armchair.

Suppose that we are curious about the size of the Earth, how could we determine that size? Your first reaction might be that it is well-nigh impossible for little old you to determine the size of the Earth. But a bold and ingenious mathematician, named Eratosthenes, did just that over 2,250 years ago. At that time, the known world was rather circumscribed (Fig. 6.4), and so Eratosthenes' feat is all the more astounding. In fact, most people of the day believed that the Earth was flat. Truth be told, the Flat Earth Society is alive and well to this day and preaches its very own special brand of ignorance on the internet.

---

[2] The actual wavelength is 12.2 cm for the oven microwaves, so if you measured 6 cm between hot spots you have done very well.

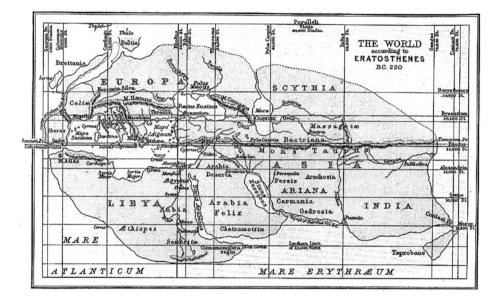

Figure 6.4: Facsimile of a map of the world at the time of Eratosthenes. (Image courtesy of FCIT: https://etc.usf.edu/maps)

If we follow in Eratosthenes' footsteps, we can figure out the size of the Earth with essentially just a stick in the ground. He noticed that, in the city of Syene (now Aswan, Egypt) at noon at the time of the Summer Solstice[3], the Sun's rays would shine directly down a well; that is, they would make no shadow as the rays were directly overhead. At the same time, Eratosthenes, who lived in Alexandria in Egypt, measured the shadow that the Sun's rays made using a rod placed in the ground, which came out at 7.2° as in Fig. 6.5[4].

As Eratosthenes was aware of Euclid's *Elements*, he knew that the angle of the shadow that he measured as 7.2° would be the same angle as that made by a line through the stick joining the center of the Earth, measured against a line through the well in Syene with the center of the Earth[5].

As the complete circumference of the Earth is 360°, 7.2° represents 1/50*th* of that circumference. The modern-day distance between Alexandria and Syene is about 800 km, although Eratosthenes had to use the less accurate figure of 5,000 *stadia*, available at the time. Therefore, if 1/50*th* represents 800 km on the Earth's circumference, then the entire Earth is: 50 × 800 = 40,000 km around. Thus,

---

[3] Now the 21st of June, but the modern calendar was not around then.

[4] Some say Eratosthenes used a tower, or statue, but the details do not matter as long as the object was vertical.

[5] These are the equal opposite interior angles made by a line crossing two parallel lines, in this case representing extensions of the Sun's rays.

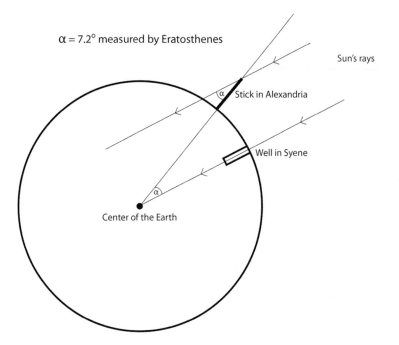

$\alpha = 7.2°$ measured by Eratosthenes

Sun's rays

Stick in Alexandria

Well in Syene

Center of the Earth

Figure 6.5: The geometry of the Sun's rays at the Summer Solstice, as noted by Eratosthenes around 240 B.C. The two angles denoted by α are equal as they are opposite interior angles made by a straight line crossing two parallel lines. Not to scale. (Illustration courtesy of Katy Metcalf.)

following Eratosthenes, we have obtained the circumference of the Earth with a fundamentally abstract geometric result of Euclid and a clever bit of reasoning[6].

Modern methods give the circumference of the Earth as 40,075 km, so we and Eratosthenes have done extraordinarily well.

HEAVENS ABOVE

Leaping ahead through the millennia, a common belief of the Middle Ages was that the Earth was the center of the Universe and that all the planets and stars revolved around it in perfect circular orbits. This was certainly the established view of the Church and seemed self-evident to anyone who looked skyward.

There had been doubters over the millennia, including Aristarchus of Samos (*ca.* 310 – *ca.* 230 B.C.), and of course the most notable was Copernicus, who

[6] Using the distance of 5,000 stadia, Eratosthenes would have obtained 25,000 stadia for the circumference of the Earth, which equals approximately 46,250 km, depending on the exact length of a stadion, which is not entirely certain.

published his *De revolutionibus orbium coelestium* (*On the Revolutions of the Heavenly Spheres*) in 1543. In the world of Copernicus, the Sun was the center of our Solar System and the Earth and planets revolved around it (although he still thought they traveled in perfect spheres around the Sun, which he got wrong). This view was supported by Galileo who, using the world's first telescope, observed that Jupiter had its own mini-solar system of moons. Galileo saw the four moons that were mentioned previously.

A key step in understanding the Universe was taken by Johannes Kepler, drawing on the data collected by Danish astronomer Tycho Brahe. Kepler empirically derived three laws that govern orbiting bodies, which involve the geometric shape of an *ellipse*[7].

By placing two stakes in the ground with a loose piece of string tied between them, an ellipse can be swept out by pulling the string taut with a third stake and moving it along the ground 360 degrees (Fig. 6.6). The two points represented by the two stakes in the ground are called *foci*. Thus, an ellipse will have the property that the sum of the two distances from any point on the curve to the two foci remains constant.

The basic features of an ellipse (centered at the origin in the plane) are the two foci (located at the points $(c, 0)$ and $(-c, 0)$, the lengths of the *semi-major axis a*, and *semi-minor axis b*, and the *eccentricity* $e = c/a$. This latter measures the degree to which the ellipse figure differs from a circle (which has eccentricity zero). Indeed, a circle is a special case of an ellipse for which the two foci are the same point, namely the center. The Earth's orbit around the Sun is nearly circular and has a very low eccentricity value of 0.0167, whereas the planet Mercury has a more highly elliptical orbit with an eccentricity of 0.21.

Kepler's 1st Law: All planets move in elliptical orbits with the Sun at one focus (see Fig. 6.6).

Kepler's 2nd Law: All planets sweep out equal areas with the Sun in equal times (see Fig. 6.7).

Kepler's 3rd Law: The square of the period of any planet is proportional to the cube of the semi-major axis.

In his *Principia Mathematica*, published in 1687, Isaac Newton was able to derive Kepler's Laws mathematically from his own Laws of Motion and Law of Universal Gravitation (see Chapter 7). The latter was not derived by Newton from empirical data, but was an axiomatic law that applied to all bodies in the Universe, from humans to galaxies. Hence the use of the word 'universal'.

Using the mathematical laws that govern planetary bodies, we can even run our understanding of the Solar System for thousands of years into the future or thousands of years into the past, to see where each planet will be, or was, at a given moment. One complicating factor is the gravitational effect that the planets have on each other, and this must also be taken into account in addition to the

---

[7] The first two were published in 1609 (the year the telescope was invented) and the third in 1619.

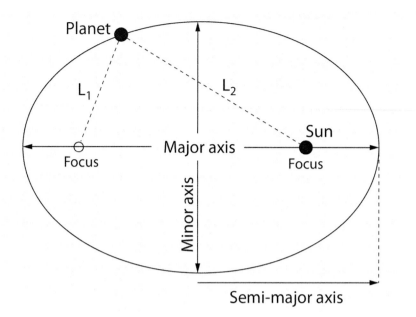

Figure 6.6: The property defining an ellipse is that the sum of the distances, $L_1 + L_2$, is a constant value. (Illustration courtesy of Katy Metcalf.)

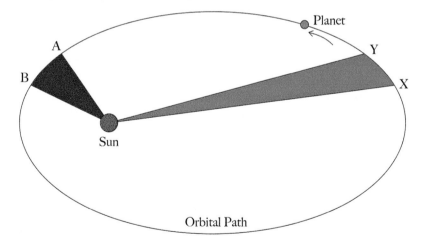

Figure 6.7: An illustration of the second of Kepler's laws that an orbiting body sweeps out equal areas in equal time regardless of where it is on its elliptical orbit around the Sun. Kepler deduced this as a consequence of his observation that planets closer to the Sun travel faster than those further out. The same law holds true for all orbiting bodies in the Universe. (Illustration courtesy of Katy Metcalf.)

Sun. All of this scientific understanding, and much more as we shall see, has been painstakingly accumulated since mankind first look skyward and wondered what it was all about.

Indeed, subsequent to the discovery of the planet Uranus in 1781 by William Herschel, it was noticed that the planet's orbit exhibited small deviations, as if perturbed by the gravity of some, as yet unknown, more distant planet. Finally, during 1845–1846, both the French mathematician Urbain Le Verrier and English mathematician John Couch Adams made mathematical calculations which not only explained the discrepancies in the orbit of Uranus, but also predicted the location in the heavens where the new planet should be found. On the night of September 23–24, 1846, the new planet (Neptune) was indeed found by the Berlin Observatory, working from Le Verrier's calculations.

But the matter did not end there, as there was a lot of national pride between France and England at stake concerning the priority of the discovery. This issue has been much studied by historians of Science, and in the modern view it is fair to say that the nod should go to Le Verrier for the discovery of the planet Neptune[8], along with Mathematics of course. In addition, credit must also be given to the person who peered through the telescope in Berlin, Johann Gottfried Galle.

But our understanding of what is up there is not limited to our local Solar System. In fact, most astronomers do not even bother with the planets any more, as they have the entire Universe at hand to study.

In his book *The Course of Positive Philosophy* (1842), French philosopher Auguste Compte stated that one thing Man would never know was the composition of a star. This certainly seemed a reasonable assumption at the time it was uttered. But it was only years later that, by carefully examining the light that comes from stars (and our Sun), it became possible to determine that the vast majority of stars are mostly composed of hydrogen gas (about 90%), some helium (about 10%), and a smattering of various heavier elements. In fact, at least 67 different elements have been identified in our nearest star, the Sun. There may even be more, but their abundance is below the level of detection.

So now we know what stars are made of, and what is more, we know the temperature of any particular star and even how heavy they are. What would Compte make of all that, one wonders? We also have a good idea of how far away they are too, usually calculated using a measurement called *light-years*.

LIGHT-YEARS

When it comes to the Universe, the scale is almost beyond human comprehension. Distances are often measured in *light-years*, which is the distance that a beam of light would travel in one year. Considering that a beam of light travels 300,000 km

---

[8] The title speaks volumes: W. Sheehan; N. Kollerstrom; C. B. Waff, The Case of the Pilfered Planet – Did the British steal Neptune? *Sci. Amer.*, Vol. 291, pp. 92–99, 2004.

in one second, we find that in an entire year it will have travelled roughly 10 trillion kilometers (6 trillion miles). That is the distance of one light-year.

The galaxy named Andromeda is some 2.5 *million* light-years away − and that is one of our closest neighbors. In other words, the light that hits our eyes when we look at the Andromeda galaxy at this very moment in time actually left town 2 1/2 million years ago. Thus, we are seeing Andromeda as it was at that time. So, the Universe is really a time-machine, which has the interesting feature in that we look back in time whenever we peer into the heavens.

In fact, there is a galaxy called IOK-1, lying at a distance of 12.88 billion light-years away. This means that it formed roughly 900 million years after the Big Bang, which is when our Universe was just in its infancy, at around 7% of its current age (Fig. 6.8).

The remarkable thing about this discovery, made in 2006, is that not only did it take nearly 13 billion years for the light from IOK-1 to reach us, but also that we had developed the technical means to pick up its incredibly faint light signals. With a bit of Mathematics, we are able to calculate such incredible distances, as will be seen in what follows.

Figure 6.8: The very distant galaxy known as IOK-1, discovered in 2006 by astronomers using the Subaru telescope in Hawaii. This is a look back into the early Universe, roughly 900 million years after the Big Bang. The galaxy appears red as its light is highly shifted towards the red end of the visible spectrum due to the expansion of the Universe. (Image courtesy of the National Astronomical Observatory of Japan.)

THE GREAT RECESSION

One rather interesting feature about our Universe is that it is expanding, and as a consequence, nearly everything in the Universe is receding from us. The actual rate of this recession is a very important quantity to know. One early attempt was published by American astronomer Edwin Hubble (1864–1934) in 1929[9]. Using observational data from Vesto Slipher and Henrietta Leavitt, coupled with his own observations, Hubble found that the radial[10] velocity of recession, v, of a galaxy at a distance d from us, is given by the famous linear relation (*Hubble's Law*),

$$v = H_0 d, \tag{22}$$

where H_0 is a constant, known as the *Hubble constant* (see Fig. 6.9). What this says is that galaxies that are further away are receding faster than ones that are closer.

The constant H_0 is one of the most important in all of Science, and its value determines the rate of expansion of the Universe. Over the years, many scientific methods have been developed to determine the value of H_0, and the value determined by each experiment has varied from 67.7 to 74 km/sec/Mpc. Here, Mpc stands for the distance *megaparsec*, meaning 1 million parsecs, which is 3.26 million light-years[11]. The discrepancy in values for H_0 will be touched upon later in the text.

To visualize why more distant galaxies are receding faster than closer ones, take a fat rubber band and, from the left end which we can denote as the origin, mark off positions at 1 *cm*, 2 *cm*, 3 *cm*, 4 *cm*. The units do not matter, so inches are fine too. Even without a rubber band, it is easy enough to do this experiment with your mind's eye. Now, place the rubber band next to a ruler, fixing the left end of the rubber band at the zero point of the ruler. Next, pull on the right-hand end of rubber band just enough so that the 1 *cm* mark on the rubber band moves along to the 2 *cm* mark on the ruler. That is all you need to do.

What happens to the mark that was at 2 *cm* on the rubber band? It will now be located at 4 *cm* along the ruler; the 3 *cm* mark on the rubber band will be positioned at 6 *cm* on the ruler; and 4 *cm* on the rubber band will be at 8 *cm* on the ruler. So far so good.

Suppose now that the stretching of the rubber band takes 1 second. This means that the first point on the rubber band, having moved from its initial position of 1 *cm* to its new one at 2 *cm*, has travelled at a speed of 1 *cm/sec*. The point that was at 2 *cm* on the rubber band (having expanded to 4 *cm*) will have travelled at 2 *cm/sec* from the left end point. Similarly, the point at 3 *cm* will have travelled at 3 *cm/sec* and the point at 4 *cm* will have travelled at 4 *cm/sec* from the fixed left-hand end.

---

[9] It will be noted later in the text that Edwin Hubble was not the first to uncover the expansion of the Universe. Moreover, the work of Vesto Slipher on the redshifts of galaxies, and that of Henrietta Leavitt on connecting Cepheid variable stars (see later in this section) with distance, were both instrumental in the formulation of Hubble's Law.

[10] The radial velocity is the line of sight velocity from Earth.

[11] The parsec, rather than light-years, is a favored unit of measurement with astronomers for technical reasons. We will generally stick with light-years.

Now, imagine that space is like the rubber band. We are the observers positioned at its left-hand end, and the points on it are other galaxies. From the above experiment, it now makes sense how the galaxies further away are receding faster from us as space itself expands.

More specifically, the formula in eq. (22) is saying is that for each Mpc further in distance, the rate of the expansion of space is increasing (taking a reasonable average) by 70 km/sec[12].

It should also be noted that the Hubble constant has not actually been constant over the age of the Universe. It was larger in the past and has been decreasing with time. Therefore, the subscript zero in H_0 indicates its value at the present time. It should really have been called the 'Hubble parameter' as it varies with time. A more complete discussion can be found in Appendix XVI but requires a passing acquaintance with the notion of the 'rate of change', i.e. the derivative.

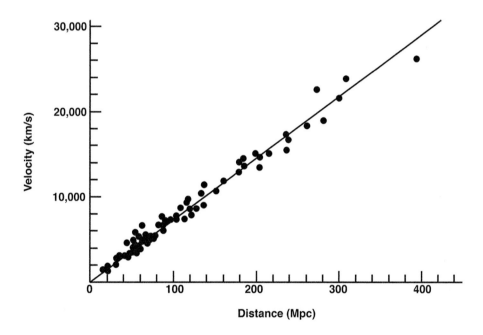

Figure 6.9: Graph showing that the radial velocity of recession from us is a constant multiple of the distance. The slope of the straight line is the Hubble constant, giving a value of $H_0 = 73$ km/sec/Mpc. (Illustration courtesy of ASU and Katy Metcalf.)

In practice, the velocity of a distant galaxy can be determined by examining its light with an instrument called a spectroscope and determining its redshift; that is, how much the light has shifted towards the red end of the spectrum. This so-called

---

[12]The Andromeda galaxy is actually moving towards us, as it has its own intrinsic motion, so that its light is shifted toward the blue end of the spectrum. In about four billion years, the Andromeda galaxy and our Milky Way galaxy will come together in a cosmic embrace.

redshift is denoted by the variable, z,[13] and the line-of-sight (radial) velocity of recession from us is related to the redshift by the simple formula,

$$v = cz,$$

where c is again the velocity of light[14].

So, for example, if the redshift of a galaxy is measured to be $z = 0.00935$, then the recessional velocity is $v = 300{,}000 \, km/sec \times 0.00935 = 2{,}805 \, km/sec$. Compare that speed with that of a typical rifle bullet, which is about $1.3 \, km/sec$, and we can see that this galaxy is receding from us more than 2,000 times faster than that speeding bullet. Converting the velocity to distance via Hubble's Law gives

$$d = \frac{v}{H_0} = \frac{2805}{70} = 40 \, Mpc.$$

Converting parsecs into something more familiar gives a distance of about 130 million light-years. Actually, we have just computed the distance to the galaxy NGC 3147 (Fig. 6.10), using its known redshift data[15]. Astronomers would consider NGC 3147 as a relatively nearby galaxy (by comparison, see Fig. 6.14 later in the chapter).

We should remark here that there are various other methods for determining distances in the Universe that have nothing to do with the redshift. In fact, there is a complete 'distance ladder' of techniques, with one leap frogging ever further out leading to the next.

For instance, the nearest stars will appear to change their position (known as *parallax*) relative to distant background stars as the Earth takes up opposite positions in its orbit around the Sun. Holding up a finger in front of your face and alternately closing your left and then right eye illustrates the same phenomenon. Elementary trigonometry then provides the star's distance. This is the first rung, so to speak, of our distance ladder.

Indeed, the parallax angle α in Fig. 6.11 is measured so that the distance to the nearby star is given by the simple expression:

$$\frac{r}{D} = \tan \alpha.$$

We can then solve for D using the known radius r of the Earth's orbit.

---

[13] If λ_0 is the wavelength of the unshifted light, and λ_1 is the redshifted wavelength (measured by looking at the light's spectrum), then the redshift is given by: $z = \frac{\lambda_1 - \lambda_0}{\lambda_0}$. This just denotes how much the light has shifted compared with its original wavelength.

[14] This simple formula works for redshift values less than about $z = 0.3$.

[15] From NASA/IPAC Extragalactic Database (NED).

Figure 6.10: The beautiful spiral galaxy NGC 3147 with its conspicuous dust lanes and spiral arms full of hot, young, blue stars. It is roughly the size of our Milky Way galaxy. The light seen in this image left the galaxy about 130 million years ago. Surely this galaxy contains other civilizations that are right now looking at us? (Image courtesy of ESA/Hubble & NASA, A. Riess *et al.*)

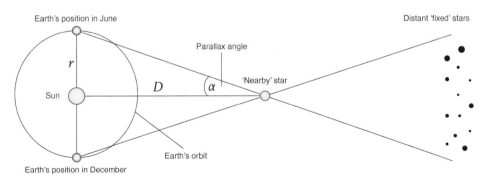

Figure 6.11: Measuring the parallax angle α to nearby stars allows for the simple determination of their distance by trigonometry. (Illustration courtesy of Katy Metcalf, reprinted with permission of IOP Science.)

The next rung on the ladder has to do with so-called Cepheid variable stars[16]. These are large pulsating stars which exhibit regular variations in their brightness and have been detected in many galaxies. The length of time between brightness peaks (much like the wavelength measurement in Fig. 6.1), known as the star's *period*, turns out to be dependent on the star's intrinsic brightness. This crucial fact was discovered in 1912 by American astronomer Henrietta Swan Leavitt (1868–1921).

In other words, stars that have longer periods are intrinsically brighter. Since the periods can vary over an interval of days to months, they are very amenable to observation by astronomers.

The upshot of this fortuitous celestial circumstance is that if we observe two Cepheids having the same period (hence intrinsic brightness), but one appears brighter than the other, then we know that the dimmer one *must* be further away. In order to place these relative distances onto a proper distance scale, several nearby Cepheids had their parallax measured. This revealed their actual distances from Earth and has allowed all the other Cepheid distances to be calibrated in terms of their corresponding periods.

The Mathematics involved is all elementary, and an example of how the distance to a galaxy containing a Cepheid variable can be calculated is found in Appendix XIV.

Sometimes galaxies are just too distant to make out individual stars like Cepheids. In this case, another distance measuring technique is used which involves a phenomenon that can be seen virtually across the entire known Universe, but is not as common as Cepheids. This is a type of supernova[17] explosion (called Type Ia) whose mechanism is well understood. By comparing the observed peak brightness with the expected intrinsic peak brightness, the distance to the supernova can again be determined.

Cepheids and Type Ia supernovae are just two approaches of many that are known as 'standard candles', in the sense that there is some intrinsic brightness – like that of an ordinary candle – that can be compared to its observed brightness[18]. If a candle's light is very faint, we know that the candle is very distant, whereas a candle close up will appear much brighter. Indeed, a common measurement of illumination is the *foot-candle*: the intensity of a candle's light on a surface at a distance of 1 foot. See the discussion in Chapter 7 involving the inverse square and Fig. 7.8 for just how light intensity naturally varies with distance.

---

[16]The name derives from the classical Delta Cephei variable star, whose period is 5.366249 days.

[17]A supernova is an exploding star that has come to the end of its life cycle.

[18]Another standard candle has the colorful name, Tip of the Red Giant Branch (TRGB), due to the ignition of helium in the star's core and its consequent flareup at a standard absolute magnitude. Our Sun will experience this phase.

Astronomers have just made this notion mathematically explicit in their calculations using standard candles (see eq. 40).

All the various distance techniques are used in order to determine the value of Hubble's constant, H_0. While all the values lie in the same ballpark, there is a disturbing discrepancy, as we shall now see.

THE UNIVERSE IS FLAT

If you draw two lines on a sheet of paper that are parallel, then of course they will not meet, since that is what is meant by being parallel. But what if, in your imagination, you extend the two lines across the Universe? Will they still never meet, or is the Universe in its largest extent somehow curved and the lines will eventually cross or even diverge?

It turns out that the theoretical crucial number in question depends on how much stuff there is in the Universe and is called the *critical density*, ρ_c. This vital number is given by the formula

$$\rho_c = \frac{3H_0^2}{8\pi G},$$

where H_0 is our new friend the Hubble constant, and G is Newton's gravitational constant[19]. Of course, the value heavily depends on what value is taken for the Hubble constant, especially as it is squared. Taking a value of H_0 = 70 km/sec/Mpc, and doing the math, yields

$$\rho_c = 9.2 \times 10^{-30}\, \frac{gm}{cm^3} \approx 10^{-29}\, \frac{gm}{cm^3},$$

which is exceeding small; roughly the equivalent to the mass of six hydrogen atoms on average in every cubic meter of space[20].

---

[19] This equation for the critical density derives from a solution to Einstein's field equations (eq. 30), given by Alexander Friedmann (see Chapter 7) for a Euclidean (flat) Universe, which can be written in the form:

$$H^2 - \frac{8\pi G}{3}\rho = 0,$$

where H is the (variable) Hubble parameter from eq. (43) and ρ is the mass-energy density in the Universe. Solving for ρ gives the critical density.

[20] The mass of a hydrogen atom is 1.67×10^{-24} grams. The critical density can also be expressed as

$$\rho_c = 9.2 \times 10^{-24}\, \frac{gm}{m^3} \approx 10^{-23}\, \frac{gm}{m^3}.$$

The critical value is only approximate at the present time, as the value of the Hubble constant is not known exactly. Since space is curved by mass, according to the Theory of Relativity (discussed in the next chapter), the amount of matter in the Universe determines its overall geometry.

Once the density has been scientifically measured, there are three possible outcomes: the density is higher, lower, or equal to the critical density ρ_c. This aspect is commonly expressed by a single *density parameter* Ω_0:

$$\Omega_0 = \frac{actual\ measured\ density}{\rho_c}.$$

Thus, if the measured density is higher than the critical density, then $\Omega_0 > 1$. If the measured density is lower than ρ_c, then $\Omega_0 < 1$. If they are both equal, then $\Omega_0 = 1$.

In the case that the matter in the Universe is higher than the critical density, that is, $\Omega_0 > 1$, then the Universe is said to be *closed* and finite, so that initially parallel rays of light follow curved paths that will eventually meet, like on the surface of a sphere (Fig. 6.12 left).

If the mass density of the Universe is lower than the critical density, that is, $\Omega_0 < 1$, the Universe will be *open*. Lines starting out as parallel light rays diverge in such a Universe as the overall shape is hyperbolic (Fig. 6.12 center).

In between these two scenarios is a Goldilocks Universe that is called 'flat', where the geometry is Euclidean and parallel light rays neither meet nor diverge (Fig. 6.12 right).

$\Omega_0 > 1$ $\Omega_0 < 1$ $\Omega_0 = 1$

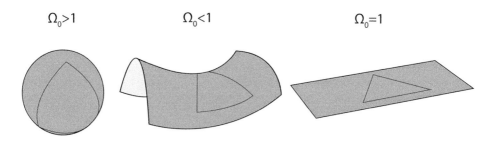

Figure 6.12: The geometries of the Universe: spherical (angles of a triangle $> 180°$), hyperbolic (angles of a triangle $< 180°$), or flat (angles of a triangle $= 180°$), depending on whether there is a high, low, or just-right density of matter, respectively. See also Fig. 1.7 in Chapter 1. (Illustration courtesy of Katy Metcalf.)

So, what is the verdict? Two long-term studies, WMAP[21] and subsequent refinements by the Planck Space Telescope (operated by the European Space Agency

---

[21] Wilkinson Microwave Anisotropy Probe launched by NASA in 2001, with the data analysis published in 2012.

from 2009–2013)[22], found that the density of matter in the Universe, including ordinary matter (4.9%), dark matter (26.8%)[23] and dark energy (68.3%)[24], is virtually that of the critical density, so that $\Omega_0 = 1$, and as a consequence the geometry of the Universe is flat up to a high degree of accuracy (margin of error is 0.4%).

The WMAP and Planck results are in very close agreement with each other, both giving a Hubble constant at the 67–69 km/sec/Mpc mark. Both WMAP and Planck measured temperature variations across the sky in the cosmic microwave background radiation, which is the temperature remnant left over from the Big Bang.

But astronomical observations employing a variety of methods, including those from the Hubble Space Telescope, give values of the Hubble constant around the 72–76 km/sec/Mpc mark[25]. The two scientific approaches have yet to be reconciled and the discrepancy in Hubble constant values is a major concern; some would even say a 'crisis' in Cosmology.

Although the Universe is apparently flat in the geometric sense above, we still do not know about the *topology* of the Universe; that is, its overall shape. Draw some initially parallel lines on a donut (before you eat it), and you will see that they remain parallel, so the two-dimensional donut's surface (mathematically known as a *torus*), is geometrically flat. So is the geometry on the surface of a cylinder, as well as on that of a Möbius strip.

The most natural – and arguably most likely – shape of the Universe is just an ordinary three-dimensional Euclidean space that is infinite in extent (no boundaries). But such an infinite space is not really the most economical to create. Invoking some sort of cosmic 'principle of least action' (see Chapter 7), it is a lot less troublesome to craft a finite Universe. Just such a proposal was made in 1984 by Russian scientists Alexei Starobinsky and Yakov Zel'dovich, and their model was the surface of a three-dimensional torus (donut). Of course, such a shape is impossible to visualize, but it can be described mathematically[26]. Such a Universe has no boundaries, but is finite in extent.

Now that is something to chew on.

---

[22] Planck also found that the age of the Universe was 13.8 billion years old, and the value of the Hubble constant was determined to be 67.4 km/sec/Mpc.

[23] This is an as yet unknown form of matter, whose effects are observed as an excess of gravitational mass found in galaxies and clusters of galaxies that is not the result of visible matter in the form of stars, gas, and dust.

[24] An unknown form of energy with a repulsive force that can explain the apparent accelerating expansion of the Universe. It is thought that it is the result of the vacuum energy which will be discussed in Chapter 8.

[25] Recent astronomical studies go by such colorful names as SH0ES and H0LiCOW. Other studies such as TRGB have pegged the value of H_0 at ~70, right in the middle of the controversy.

[26] The surface of an ordinary donut is two-dimensional, since a point on the surface can only move in two perpendicular directions.

MEASURING THE INVISIBLE: BLACK HOLES

One thing common to most galaxies is that they have a supermassive black hole at their center. A black hole is an object possessing such intense gravity that nothing, not even light, can escape beyond its surface[27]. In 1916, just such an entity was found by the German physicist/astronomer Karl Schwarzschild to be a theoretical consequence of the recently published General Theory of Relativity, although even Einstein himself did not believe they could actually exist.

In fact, if the mass of any object were shrunk down to that of a sufficiently small sphere, it would become a black hole. The radius of that sphere, R_S, is known as the *Schwarzschild radius* and can be expressed by the simple formula[28]

$$R_S = 3\frac{M_{BH}}{M_S}$$

where M_{BH} is the mass of the black hole and M_S is the mass of the Sun, with the result given in kilometers. Therefore, for the Sun to become a black hole it would have to shrink down to a radius of 3 *km*.

On the other hand, for the author to become a black hole, the preceding equation gives his personal Schwarzschild radius as, $R_S = 1.05 \times 10^{-25}$ *m*. By contrast, the radius of an electron is 2.82×10^{-15} *m*. The reader's Schwarzschild radius will be much the same as that of the author since in either case we must divide by the mass of the Sun. Therefore, neither of us is ever likely to turn into a black hole.

Although the concept of a black hole seems very modern, it was actually considered by two scientists in the 18th century, in particular Englishman John Michell and Frenchman Pierre-Simon Laplace. But they remained idle curiosities, even after the 20th century mathematical work of Schwarzschild.

Things began to change in 1963, when New Zealand mathematician Roy Kerr found a solution to Einstein's relativity equations (eq. 30) that described a rotating black hole, and a year later the first black hole candidate was discovered, now known as Cygnus X-1.

Black holes are very mysterious objects and not well understood. They have what is called an 'event horizon', which is the spherical shell boundary inside of which nothing can escape the intense gravity of the black hole.

---

[27] The term was first proposed by American astronomer John Wheeler in 1971.

[28] The formula found by Schwarzschild was $R_S = \dfrac{2GM}{c^2}$, where G is the gravitational constant, M the mass of the object, and c the velocity of light. The version given in the text is a little more user friendly. Schwarzschild was the first to solve Einstein's field equations (eq. 30) for a non-rotating black hole and his eponymous radius appears in that work.

Although black holes are now very much a part of the cosmological landscape, the devil is in many of the details. In a question posed by physicist philosopher Kurt Curiel to many scientists working in the field, on 'What is a black hole', he received a multiplicity of replies[29]. Nevertheless, he concludes by saying, "I would rather say that there is a rough, nebulous concept of a black hole shared across physics, that one can explicate that idea by articulating a more or less precise definition that captures in a clear way many important features of the nebulous idea, and that this can be done in many different ways, each appropriate for different theoretical, observational, and foundational contexts."

The Cygnus X-1 black hole, about 6,070 light-years distant, is accreting matter off a hot blue supergiant companion star. This matter spirals into the black hole, forming a thin, hot disk which gives off intense X-ray emissions that can be detected on Earth.

The mass of the Cygnus X-1 black hole has been measured at 14.8 times the mass of the Sun. If we put this mass into the preceding Schwarzchild radius formula,

$$R_S = 3\frac{M_{BH}}{M_S} = 3 \times 14.8 = 44.4 \, km.$$

Moreover, a point on the event horizon is rotating at a staggering 800 revolutions per second. Compare that with the hard drive in your average desktop computer, which spins at a rate of 120 revolutions per second (= 7,200 RPM) and which is also vastly smaller[30].

Black holes come in various sizes, given by the radius of the event horizon, as well as masses. A *stellar mass* black hole forms when a suitable massive star collapses in its old age. They typically are a few tens of solar masses[31], whereas a *supermassive* black hole can be hundreds of thousands to billions of solar masses.

A supermassive black hole lies at the heart of our own Milky Way galaxy and has a mass of 4.3 million solar masses. This figure can be deduced by examining the orbits of stars near the center of our galaxy and seeing how the orbits are perturbed. Indeed, it was an idea proposed by Laplace, who suggested that an invisible black hole could be inferred from a study of the orbits of stars around it. Moreover, a similar idea can be applied to galaxies even millions of light-years distant, in spite of the fact that the supermassive black hole at its core is itself invisible. This is a rather extraordinary feat (see Fig. 6.13).

---

[29] E. Curiel, The many definitions of a black hole, *Nature Astronomy*, 3, 27–34, 2019.

[30] A point at the equator of the rotating event horizon is traveling a distance given by $C = 2\pi r$, where $r \approx 44km$, the Schwarzchild radius. Therefore, in one second the point has travelled a distance of $800 \times 2\pi r = 221,168 \, km$. This means that the point on the event horizon is traveling at 74% of the speed of light (300,000 km/sec). Moreover, as the event horizon bulges at the equator to some extent, the velocity there is even higher.

[31] That is, times the mass of our Sun. *Solar masses* is a convenient measuring unit of mass in Astronomy.

There is thought to be a supermassive black hole lurking at the center of most large galaxies and each will cause the stars in its near vicinity to orbit rapidly around it. The velocity of these stars can be measured by their redshift value, taking into account the velocity of the galaxy as a whole, of course. Once a handful of velocity values have been determined, the average velocity is calculated. The larger the black hole mass, the more the velocities will deviate from this average value and the black hole mass can be computed as a consequence of Newton's laws.

The amount of this variation in the velocity values from the average is just the *standard deviation* that we have discussed in Chapter 2. You may recall that a small standard deviation indicates that the values are not too widely dispersed from the average (mean), and a larger standard deviation indicates a wide spread of values.

However, with the exception of nearby galaxies, it is not possible to track the motion of stars that are sufficiently close to the supermassive black hole to be affected by its gravity. Even the Hubble Space Telescope does not have the required resolution. So, in the case of distant galaxies, what can be measured is the velocity dispersion about the mean of stars much further out in the galaxy, typically a few thousand light-years distant from the black hole at the center.

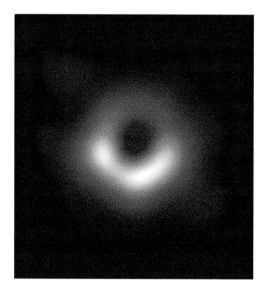

Figure 6.13:   This is the first image of a black hole taken in 2019 by the Event Horizon Telescope array of radio telescopes, situated across the globe and all synchronized by atomic clocks. This supermassive Kerr rotating black hole, some 38 billion kilometers in diameter, lies at the heart of the supergiant elliptical galaxy Messier 87 and is surrounded by a rapidly rotating ring of glowing gas and dust. (Image courtesy of Event Horizon Telescope.)

Denoting *this* standard deviation by σ then the mass of the black hole can be computed by the empirically-derived straight-line relationship:

$$\log\frac{M_{BH}}{M_S} = 4.36\log\left(\frac{\sigma}{200\,km/sec}\right) + 8.32, \tag{24}$$

where M_{BH} is the mass of the black hole and M_S is the mass of our Sun[32].

In the case of the galaxy M87, for the black hole imaged in 2019 in Fig. 6.13, the value of σ is 440 km/sec, so that doing the math gives: M_{BH} = 6.5 billion times the mass of our Sun.

We have just calculated the extraordinarily large mass of something unseen that lies at the core of a distant galaxy. It is hoped that the reader is duly impressed with this achievement.

So, just how far away is it? For that we just need the redshift, which has a value of z = 0.00428. Using our recessional velocity formula, $v = cz$, we find that the galaxy is receding from us at whopping speed of 1,284 km/sec (2.87 million mph). Using the Hubble formula, $v = H_0 d$, the distance works out to 18.3 million parsecs, which equates to about 60 million light-years[33].

We could say that our image in Fig. 6.13 is 60 million years out of date, and this would be true, which means that 'now' − that is, 'right now, Earth time' − the galaxy is moving even *faster* and is now *further* away from us. But in order to avoid total confusion, astronomers think of 'now' in the sense of what the data says, and in this case, the data has taken 60 million years to get here. This is also known as the 'lookback time'.

Originally, black holes were thought to have no entropy. Recall from Chapter 4 that the Second Law of Thermodynamics says that entropy of a closed system cannot decrease. Now a hot gas is highly disordered and so has a lot of entropy, and if it gets sucked into a black hole then entropy cannot simply become lost once the black hole settles down. This apparent violation of the Second Law was pointed out by astrophysicist Jacob Beckenstein, who proposed that a black hole's entropy was proportional to the surface area of its event horizon. However, this idea was initially opposed by the famous English physicist, Stephen Hawking.

---

[32] Equation modified by the author from: S. Tremaine, *et al.*, The slope of the black hole mass versus velocity dispersion correlation, *Astrophys. J.* Vol. 574, pp. 740−753, 2002.

[33] Here we have taken the Hubble constant to be 70 km/sec/Mpc. Some sources give the distance as 53−54 million light-years, but our determination is good enough via our simple approach and other sources agree with it.

The upshot is that a black hole does indeed have entropy. Through the work of Beckenstein and Hawking, the entropy of a black hole, S_{BH}, is given by the formula (setting all constants involved equal to 1 for simplicity),

$$S_{BH} = \frac{A}{4},$$

and A is the surface area of the event horizon[34]. Moreover, the information about that entropy is contained in the event horizon.

In 1993, this led physicist Gerard 't Hooft[35] to propose the Holographic Principle mentioned in Chapter 1, in which all of the Physics that takes place in a region of space can be described by information that occurs on the boundary of the region's surface. The idea was popularized by physicist Leonard Susskind, and has been enthusiastically explored by many physicists since.

A GALAXY FAR, FAR, AWAY

As will be seen in the next chapter, one aspect of the Mathematics of the General Theory of Relativity is that a massive object like our Sun not only slows down time as you get closer to it, due to its gravity, but that its gravity also bends the light that passes near it.

The same thing happens with a galaxy or cluster of galaxies, which can bend the light from whatever lies in the distance behind it. Depending on the distances involved and the particular alignments, magnification and multiple images, as well as arcs, or rings, can be produced by an intervening galaxy, in a phenomenon called *gravitational lensing*. This magnification effect has been used to find extra-solar planets, as well as to peer into the very early origins of our Universe, which began some 13.8 billion years ago.

One such example is the discovery of galaxy SPT0615-JD that formed in the early evolution of the Universe. It was detected by the gravitational lensing effect of a cluster of galaxies which multiplied the size and brightness of the distant galaxy (Fig. 6.14).

---

[34] The full formula is: $S_{BH} = \frac{kc^3 A}{4G\hbar}$, where k is the Boltzmann constant that comes from the study of entropy, c is the velocity of light, G the gravitational constant, and $\hbar = \frac{h}{2\pi}$, for Planck's constant h.

[35] Gerard 't Hooft shared the 1999 Nobel Prize in Physics with Martinus Veltman.

The distance of the remote galaxy is again worked out from the redshift, which is $z = 9.9$, but the formula used previously, $v = cz$, no longer applies as the situation is more complex[36]. But the image of SPT0615-JD is from 13.3 billion years ago, just 500 million years after the Big Bang. Interestingly, it did not take long, relatively speaking, for galaxies to begin forming in the early Universe.

Figure 6.14: The beautiful cluster of galaxies SPT-CL J0615-5746 in the foreground, whose gravity has magnified and stretched the embryonic galaxy SPT0615-JD that is being viewed as it was some 500 million years after the formation of the Universe. (Image courtesy of NASA/ESA/STScI/B. Salmon.)

---

[36] In fact, for a redshift more than about $z = 0.3$, different Mathematics must be used.

7

The Unreality of Reality

> *Between the idea*
> *And the reality*
> *Between the motion*
> *And the act*
> *Falls the Shadow*
> T.S. Eliot

As children, our experience of the world is filled with a sense of awe and wonder. As adults, we lose this sense to some extent, in part because the world becomes such a familiar place that we take most things for granted. Even space travel has lost some of its license to thrill, and having new hi-tech gadgets rolling out each year to titillate our senses has become an expectation. On the other hand, most people would agree that the birth of their own child is something that is still pretty awesome.

So, while cell phones and iPads are *de rigueur* and Science, which gave us cell phones and iPads among a plethora of other electronic gadgets, seems to have most of the answers, many people are still looking for a sense of mystery in life, perhaps more than ever as Science has taken away much of that mystery. No longer is life dependent on the whims of capricious gods, whose demands must be met so that the sea does not rise up and swallow us, or the Earth rent open and devour us.

UFOs, mysticism, paranormal activity, haunted houses and ghosts provide many people with modern day mystery. While these enjoy a wide public appeal, the aim of this book is to demonstrate that life is still full of *real* mystery, full of demonstrable mystery from Science itself. On the other hand, questions such as, "Why is there a Universe at all rather than nothing?" or "Is there still meaning in

© Springer Nature Switzerland AG 2020
J. L. Schiff, *The Mathematical Universe*, Springer Praxis Books,
https://doi.org/10.1007/978-3-030-50649-0_7

life if the world is to end?" are enduring mysteries as well, but these are far too deep for the present text, and the reader is advised to look into other works considered by scientists and philosophers alike concerning these issues.

Most of us go through life thinking that we know something about the natural world that we live in. Earth, air, fire, and water were the four elements that made up the world, according to classical wisdom. But most students now learn in school that there are actually 92 naturally occurring elements, starting with the lightest ones, hydrogen, helium, and lithium... and finishing with the heaviest, uranium.

We also assume that there is an objective reality out there that we experience with our five senses, a life filled with cars, people, animals, cell phones, TVs, computers, mortgages, death and taxes. Our lives flow inexorably by, second by countless second. Well, that is the basic scenario of modern life at least, which is familiar to many of us. Yet, on another level, nothing about the real world is familiar any longer, and therein lies real mystery.

MINIATURE UNIVERSE

Let us now turn our attention away from the Universe in the large, to the same Universe but in the very, very small. A remarkable thing happened in 1928, when English theoretical physicist Paul Dirac was producing some mathematical squiggles on paper in an attempt to unite the ideas of Special Relativity with those of Quantum Mechanics. His result, now known as the Dirac equation, was a relativistic version of the Schrödinger equation we will be discussing later.

Rather surprisingly, Dirac's equation implied that there should be *anti-particles* in Nature; that is, each type of particle should have a corresponding one with the same mass but opposite physical charge. It should be mentioned that the Dirac equation and the Schrödinger equation involve the imaginary number i, yet both are used to describe the workings of our Universe at its most fundamental level.

Some four years later, in 1932, American physicist Carl Anderson demonstrated the existence of anti-electrons resulting from the high energy collisions of cosmic rays impacting the Earth's atmosphere (Fig. 7.1). *Positrons*, as Anderson called them, had the same mass as an electron but with a positive charge, whereas electrons have a negative charge.

What Dirac himself had to say on the matter some years later is interesting and is at the core of this book:

> "It seems to be one of the fundamental features of nature that fundamental physical laws are described in terms of a mathematical theory of great beauty and power, needing quite a high standard of mathematics for one to understand it. You may wonder: Why is nature constructed along these lines? One can only answer that our present knowledge seems to show that nature is so constructed. We simply have to accept it. One could perhaps describe the

situation by saying that God is a mathematician of a very high order, and He used very advanced mathematics in constructing the universe. Our feeble attempts at mathematics enable us to understand a bit of the universe, and as we proceed to develop higher and higher mathematics we can hope to understand the universe better[1]."

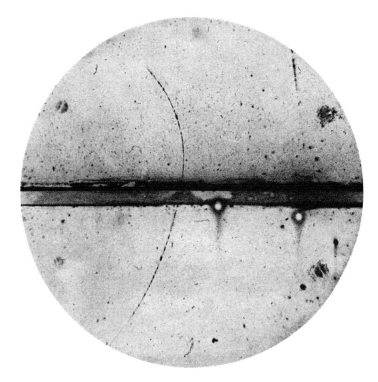

Figure 7.1: From arcane symbols on paper to reality. The particle predicted by Dirac's equation was found by physicist Carl Anderson in 1932 in this historic cloud chamber photograph. The particle's path is deflected to the left by a magnetic field, indicating that it is positively charged, not negatively charged like an electron. As it passes from bottom to top through a 6 mm thick lead plate, the particle has diminished energy and its curvature becomes even greater. (Image in public domain.)

At the age of just 31, Paul Adrien Maurice Dirac was awarded the Nobel Prize in Physics in 1933 for his stunning achievement in the prediction of the positron from his equation[2].

Indeed, it can be crudely said that Physics is just Mathematics with words.

---

[1] P.A.M. Dirac, writing in *Scientific American*, May 1963.

[2] The Nobel Prize that year was also shared with Erwin Schrödinger, for his wave equation contribution to the understanding of Quantum Mechanics.

It should also be noted that one does not need a Nobel Prize to see the trails of atomic particles in a cloud chamber. The latter can easily be made using an open plastic container with a sheet of glass to cover it, a length of weather stripping, some alcohol, and a bit of dry ice (frozen carbon dioxide).

In the old days, when the author was young, an ice cream vendor would come around the neighborhood daily, keeping the ice cream in his truck cold with dry ice. His generosity allowed the author to construct several cloud chambers and even make ill-fated attempts to bend the visible particle paths with a giant magnet. What one sees in the cloud chamber is the vapor trail of a charged particle as it passes through the supersaturated alcohol vapor produced by the dry ice placed underneath the container. Instructions can be found on the internet, and the results are impressive[3].

The positron anti-particle is also produced in certain rare radioactive decay emissions, called *beta-plus*. This type of decay occurs when a proton inside a nucleus transforms into a neutron, with the release of a positron and a particle called an electron neutrino.

When antimatter was first introduced into the theoretical framework of Physics early in the 20th century, nobody thought that this had any practical relevance. Yet today, positron emission is used in hospitals around the world on a daily basis. Positron Emission Tomography (PET) is based on the beta-plus emissions from a handful of radioisotope sources. Each radioisotope has its own specific use for seeing what is happening inside the human body. Positrons also feature in the nuclear reactions that power the Sun.

One interesting problem with anti-particles has remained unresolved since the time of the Big Bang. This singular event should have resulted in the creation of equal amounts of ordinary matter and anti-matter. If they had been created in *exactly* equal numbers, then in the first instants after the Big Bang the matter and anti-matter particles would have annihilated one another, leaving behind nothing but energy.

However, wherever one looks, from earthworms to distant galaxies, we find ordinary matter. Only about one anti-particle in a billion has persisted in the Universe. This means that there must be some very subtle asymmetry in a law of Physics, one that would not only have permitted ordinary matter to annihilate virtually all the anti-matter created, but also to go on to form the glorious Universe that we witness today. Exactly what this asymmetry is, no one yet knows. But for this slight anomaly, our Universe would be nothing more than pure energy.

---

[3] See, for example, the website: https://www.symmetrymagazine.org/article/january-2015/how-to-build-your-own-particle-detector

Since the discovery of the positron, numerous other anti-particles have subsequently been discovered in the high-speed collisions of particle accelerators. Another famous discovery based on mathematical calculations is the Higgs particle, which gives rise to the Higgs field that gives elementary particles their mass.

Equally mathematical was the discovery of *quarks* by Murray Gell-Mann, also independently discovered by George Zweig, and André Petermann in the early 1960s[4].

Quarks are elementary particles that combine in certain ways to form other particles, such as the protons and neutrons that make up the nucleus of atoms of all the elements. As fanciful as its name, quarks come in six 'flavors' ('up', 'down', 'charm', 'strange', 'top', 'bottom'), indicating differences in their charge and mass. A proton is composed of two 'up' quarks and one 'down' quark tightly bound, whereas a neutron consists of one 'up' quark and two 'down' quarks. In Gell-Mann's case, quarks came out of mathematical theory:

> "Gell-Mann's approach to the 'discovery' of quarks was highly mathematical and rather esoteric; he found that certain features of equations could be explained by treating protons and neutrons as if they were composed of triplets, but he started from the mathematical end, not from considering these triplets to be physically real particles[5]."

While the quark theory was initially received with skepticism, there is now supporting evidence from many experimental results of 'inelastic scattering' where, for example, high energy particles such as electrons are fired at protons and neutrons and the scattered remnants analyzed, like tea leaves[6].

Scattering experiments such as these (Fig. 7.2) harken back to the experiments of Nobel Prize winning New Zealand-born British scientist Ernest Rutherford in the early 20th century[7]. Atomic particles were directed at gold foil and Rutherford found that although most passed straight through the atoms of the foil, some were deflected in all directions. This scattering allowed him to work out that an atom had to possess a small, dense, positively charged nucleus at its core, containing most of the atom's mass, surrounded by mostly empty space and orbiting electrons.

---

[4] The name 'quark' was appropriated by Gell-Mann from James Joyce's *Finnegan's Wake*, and should not be confused with the dairy product of the same name. Zweig used the term 'aces' which is not nearly as poetic.

[5] J. Gribbin, p.193. See Bibliography.

[6] The Nobel Prize in Physics in 1990 was awarded jointly to Jerome I. Friedman, Henry W. Kendall and Richard E. Taylor "for their pioneering investigations concerning deep inelastic scattering of electrons on protons and bound neutrons, which have been of essential importance for the development of the quark model in particle physics." From the Nobel Prize website.

[7] The atomic particles directed at the gold foil were *alpha particles*, which consist of two protons and two neutrons bound together and are the nucleus of helium atoms. They are produced by radioactive decay, such as from radium and uranium, as well as being found in cosmic rays. Alpha particles, and the notion of a half-life of radioactive substances, were also discoveries of Rutherford.

Quarks are now a key ingredient in the Standard Model of Particle Physics. Interestingly, the mass of the Higgs particle and that of the 'top' quark seem to be linked to the ultimate fate of the Universe. However, it must be stated that there is no image of a quark, as there is in Fig. 7.1 of a positron, and that they are theoretical constructs. But the experimental evidence for their existence is very compelling.

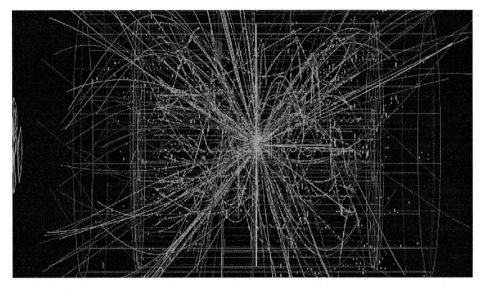

Figure 7.2: Reading the tea leaves from a high energy collision at the Large Hadron Collider between a proton and a lead nucleus. A plasma soup is created, simulating conditions immediately after the Big Bang. (Image courtesy of CERN.)

It was only in the late 19th and early 20th century that the ordinary view of the physical world took a dramatic turn in the direction of something completely mystical and nearly incomprehensible. This is of course the world of Quantum Physics, the world of atoms and sub-atomic particles.

QUANTUM WORLD

I can safely say that nobody understands quantum mechanics...
Physicist Richard Feynman, one of the founders of Quantum Mechanics.

The beginning of the 20th century saw an explosion of results coming out of the Physics governing the very small. But unlike the classical mechanics of solid bodies like bowling balls and planets, Quantum Mechanics virtually describes another universe, completely unlike the real world of ordinary experience. The quantum world appears to be in a state of suspended reality, and it is only when a measurement is made that it enters our real world.

The theory has been developed since the beginning of the 20th century in the work of numerous scientists and, like the Theory of Relativity, is completely at odds with everyday experience. The subject is very arcane and deep, and we can only mention a few highlights here.

For example, light behaves like both a particle and a wave. It all depends on how you examine light. We have already discussed about the wavelength of light and how light can be redshifted or blueshifted. This is an important fact in Astronomy. But light is also described in terms of photons, which are massless particles or *light quanta*[8].

The energy, E, of the photon only depends on its frequency and is given by

$$E = hf, \tag{25}$$

where h is the famous Planck's constant, and f is the frequency of the wave.

For example, taking a garden variety photon of yellow light that has a wavelength of 580 nanometers[9], we use eq. (21) to deduce that its frequency is

$$f = c / \lambda = 517 \times 10^{12} \ cycles \ / \ sec.$$

Therefore, the photon of yellow light has energy (putting in an appropriate value for Planck's constant)[10] of:

$$E = hf = 2.14 \, eV,$$

and an *electron volt* (eV) is a microscopic unit of energy; namely, the energy imparted to a single electron from a one-volt battery. Thus, we have calculated the energy of a single photon of yellow light, which is also a 'particle' having no mass.

Light photons from other colors, such as blue and violet light, have higher energy values as their light has higher frequencies, whereas photons of orange and red light have lower energy values as their light has lower frequencies.

The knowledge that the energy is proportional to the frequency of light is behind Einstein's explanation of the *photoelectric effect*, where light shining on, say, a metal plate, induces the release of electrons. Curiously, dimming or brightening the light has no effect on the energy of the electrons emitted, but rather the light's intensity only affects the *number* of electrons emitted. Brighter light results in the emission of a greater number of electrons from the metal plate.

---

[8] Massless in the sense of having no *rest mass*, not in the sense of converting their energy to mass via $E = mc^2$. See the subsequent photoelectric effect.

[9] By comparison, the width of a hair is 80,000 to 100,000 nm.

[10] $h = 4.136 \times 10^{-15} eV \cdot sec.$

On the other hand, the *energy* of an expelled electron only depends on the energy of the incident photon. This is akin to a billiard ball colliding with another one with its attendant transfer of energy. The photon's energy is a consequence of the light's frequency; namely $E = hf$ from eq. (25).

The photoelectric effect confirms that light not only comes in waves, but in a stream of small packets of energy which are of course our photons. In the words of the master:

> "[W]hen a light ray starting from a point is propagated, the energy is not continuously distributed over an ever-increasing volume, but it consists of a finite number of energy quanta, [photons] localized in space, which move without being divided and which can be absorbed or emitted only as a whole[11]."

At this juncture, we might ask just how much elaborate scientific equipment does one need to detect the presence of a single photon? Well, not much. Interestingly, photoreceptors in our eyes (namely the rod cells of the retina) can respond at a molecular level to a single photon. Not only that, there is also some evidence that humans can actually see the flash from a single photon, at least some of the time. Furthermore, the chance of perceiving a single photon was found to be enhanced if a previous photon was released a few seconds earlier, which is suggestive of a priming mechanism[12].

Whether light is observed in its wave form or its particle form depends on the experiment, and as it turns out, it is impossible to observe both properties at the same time[13]. Although in flight, *both* properties of light are maintained.

Now, electrons are particles that do have a mass, but they can also behave as waves. We take our cue from the case of the massless photon, which nevertheless has a *momentum*[14]. This is denoted by p (for historical reasons) and is given by the expression $p = h/\lambda$, which we rewrite as

$$\lambda = \frac{h}{p},$$

as again, h is the Planck constant.

---

[11] A. Einstein, Über einen die Erzeugung und Verwandlung des Lichtes betreffenden heuristischen Gesichtspunkt, *Ann. der Physik*, Vol. 17 (6): pp. 132–148, 1905.

[12] J.N. Tinsley *et al.*, Direct detection of a single photon by humans, *Nature Comm.*, 7, 12172, 2016.

[13] This is known as the *principle of complementarity*.

[14] This was experimentally verified by American physicist Arthur Compton, which won him the 1927 Nobel Prize. The classical definition of momentum is the tendency of a moving object to continue moving in the absence of any external force, and is given by the equation: $p = mv$; that is, mass times velocity. Therefore, a freight train traveling at 100 km/h has a lot more momentum than your car traveling at the same speed.

The *wavelength of a particle* with mass m can be described using the equation developed by French physicist Louis de Broglie (1892–1987), as

$$\lambda = \frac{h}{p} = \frac{h}{mv}, \tag{26}$$

where v is the particle's velocity, and $p = mv$ is the classical equation defining the momentum of any object with mass.

This means that an electron which does have a mass, as well as atoms and molecules, can behave like waves at the quantum scale. This is known as *wave-particle duality* and is a very strange feature of the quantum universe that in everyday experience we are oblivious to. But it is right there in eq. (26).

Much wave strangeness was developed throughout the 1920s, via a group of physicists centered around Danish physicist Niels Bohr in Copenhagen. What came out of their work was a mathematical formalism, known as the *Copenhagen interpretation*. The focus of this interpretation was based on the *Schrödinger wave equation* attributed to Austrian physicist Erwin Schrödinger, although an equivalent formulation that involved matrices had previously been proposed by Werner Heisenberg[15].

The Wave Mechanics that has prevailed until this day is a mathematical formalism that has little to do with macro-reality. For one thing, it is based on probabilities, whereas the statement, "Erwin Schrödinger is dead" requires a yes or no answer. He died in 1961.

If we take the example of a single particle such as an electron, one of its observable properties is its position, which is described by a *wave function*; that is, it obeys Schrödinger's equation concerning waves. This function, designated by, $\Psi(x, y, z, t)$, where (x, y, z) represents the electron's position in space and t is time, takes on *complex* number values. This is notable, as we are using complex numbers to describe events in the real world. Such a *state function* is able to encapsulate all that can be known about the electron. The world the function Ψ lives in is a mathematical space of vectors called a Hilbert space, which will be discussed later in the chapter.

When you take the absolute value squared of the complex number, $|\Psi(x, y, z, t)|^2$, you get a real number for each input of (x, y, z, t) that tells you the *probability* of finding the electron at that particular point in space at the time t. This result was formulated in 1926 by German physicist Max Born and is known as the *Born Rule*. You cannot really say definitively that a subatomic particle is either here or there. For a genuine electron, if you sum up all the probabilities over space, you will get the value 1; that is, it will be somewhere (Fig. 7.3).

---

[15] Matrices are just arrays of numbers having formal rules in order to do arithmetic with them.

Gone is the image of an atom with electrons whizzing about the nucleus like planets in a mini-solar system, as in the Rutherford model. In its place, there are only probabilities. When a measurement is made, say at a detector, the wave function is said to 'collapse' and the electron's location has been determined. This collapse of the wave function has always had a mystical aura about it, and still does. Suddenly, a particle goes from the realm of probability and enters the real world of detectors, tables, and chairs.

There is an alternative, rather revolutionary 'Many-Worlds' interpretation, proposed in 1956 by physicist Hugh Everett III. In this version of reality, the wave function does not collapse, but rather, during a quantum measurement, the Universe splits into parallel Universes in which each possible outcome occurs, yet the outcomes do not interact in any way with each other. Since then, as we have seen in Chapter 1, there have been various multiverse theories postulated. Indeed, Max Tegmark's Level III Multiverse is the one proposed by Everett. An interesting view of the Many-Worlds theory and the basics of Quantum Mechanics is presented in Sean Carroll's very readable account in the Bibliography.

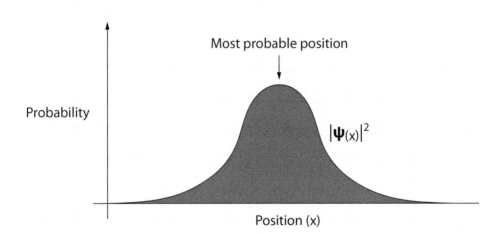

Figure 7.3: Depiction of a wave function, considering just the single x-coordinate variable, that describes the probability of finding a particle such as an electron at a particular position. Summing up all the probabilities gives the area under the curve which equals 1, as the electron exists somewhere. But its exact location is uncertain until a measurement is made. (Illustration courtesy of Katy Metcalf.)

Not everyone was happy with this situation, least of all Albert Einstein. He believed that there should be an underlying objective reality, but according to Quantum Mechanics there is no such thing. Moreover, it is not a matter of a lack of understanding of what is going on in the quantum universe, but that there is no bedrock reality to understand in the first place. Lest one should think that Quantum Mechanics is seriously flawed, it turns out to be arguably the most successful theory in all of Science. It predicts properties of particles and interactions with astonishing precision.

Indeed, in one branch of Quantum Mechanics, known as QED (Quantum Electrodynamics), a quantity known as the magnetic moment of an electron[16] has been measured experimentally to be (ignoring units):

$$1.001\ 159\ 652\ 180\ 73\ (\pm 28).$$

This compares with the best theoretical value from QED:

$$1.001\ 159\ 652\ 181\ 13\ (\pm 86),$$

an agreement to within one part in a trillion. According to theoretical physicist Daniel Styer, this is equivalent to measuring the distance from the Earth to the Moon to within an accuracy of the width of a human hair, or measuring the distance from Los Angeles to New York to within the width of a bacterium. Improvements to the above two values are still on the cards.

No phenomenon has yet been found to contradict quantum theory. The real enigma is making sense of it all.

According to author and astrophysicist Adam Becker:

"Rather than telling us a story about the quantum world that atoms and subatomic particles inhabit, the Copenhagen interpretation states that quantum physics is merely a tool for calculating the probabilities of various outcomes of experiments. According to Bohr, there isn't a story about the quantum world because 'there is no quantum world. There is only an abstract quantum physical description'. That description doesn't allow us to do more than predict probabilities for quantum events, because quantum objects don't exist in the same way as the everyday world around us. As Heisenberg put it, 'The idea of an objective real world whose smallest parts exist objectively in the same sense as stones or trees exist, independently of whether or not we observe them, is impossible'. But the results of our experiments are very real, because we create them in the process of measuring them. Jordan said when measuring the position of a subatomic particle such as an electron, 'the electron is forced to a decision. We compel it to assume a definite position; previously, it was, in general, neither here nor there… We ourselves produce the results of measurement'[17]."

---

[16] Electrons behave like little magnets, and the magnetic moment is a measure of the strength of the electron's magnetic attraction. See also footnote 24 in this chapter regarding electron spin.

[17] From A. Becker, see Bibliography.

So essentially, at the quantum level, we create a particular reality once we interfere with the system by doing a particular measurement. But prior to that there is no objective reality, only wave functions governed by probabilities. It is almost as if the world that we see and experience is no longer there once we close our eyes. To penetrate this world, words are simply not enough. The only way seems to be through Mathematics alone.

In a similar vein, there is the fantastical and not very well known *Free Will Theorem*, devised by John Conway and Simon Kochen, both of Princeton University[18]. In essence, it says that under the basic assumptions of three axioms from Physics:

If two experimenters are free to make choices about what measurements to take, then the results of the measurements cannot be determined by anything previous to the experiments.

In other words, if human beings have free will, then so do subatomic particles. It should be mentioned that the authors do not prove the existence of free will in humans. That is still a subject for debate among philosophers and scientists, but Conway and Kochen lean toward it being a reality. The author heard Conway lecture at the University of Auckland on the Free Will Theorem, and he started out by saying, "I can either pick up this piece of chalk now or not."

Moreover, the Mathematics is sound, the reasoning is very tight, and it has stood the test of time. The first two axioms are from Quantum Mechanics and are supported both in theory and experiment. The third axiom says that: "If the two experimenters are separated in space then they are able to make independent choices of measurement from one another." This is even weaker than the assumption that information cannot be propagated faster than light, from the Theory of Relativity.

The authors conclude:

"The world it presents us with is a fascinating one, in which fundamental particles are continually making their own decisions. No theory can predict exactly what these particles will do in the future for the very good reason that they may not yet have decided what this will be! ... Einstein could not bring himself to believe that 'God plays dice with the world', but perhaps we could reconcile him to the idea that 'God lets the world run free'."

While the quantum universe may seem rather abstruse, there are real world consequences. Suppose a child throws a baseball at its bedroom door. There is no chance that the baseball will end up on the other side of the door by passing through it, because the baseball lacks the energy from the child's throw to penetrate the door and it will simply bounce back. This is classical mechanics.

[18] J. Conway and S. Kochen, The Free Will Theorem, *Found. Phys.* 36, pp. 1441–1473, 2006.

But the situation is rather different at the quantum scale. Replacing the baseball with an elementary particle described by its wave function, then its location, unlike the baseball, is now determined by a probability distribution in space. This means that there is a non-zero probability that the particle will sneak through to the other side of the barrier, even though it lacks the energy to do so. In the case of a physical barrier, the thinner it is the greater the probability of passing through it. This effect was first noticed in radioactivity, where particles are able to escape from the nucleus of certain atoms. Normally, the protons and neutrons are tightly bound in the nucleus and its constituents are not able to escape.

The above is a process known as *quantum tunneling*, and it features in the nuclear fusion that powers stars, in transistors that power computers, and even in Biology. Indeed, there is a burgeoning new field of Quantum Biology that is attempting to explain various biological processes via quantum mechanical means such as tunneling. Quantum effects could also be involved in photosynthesis, or in how birds are able to navigate for long distances, which has always been a bit baffling. For more on this fascinating new field, see the excellent book *Life on the Edge* by J. McFadden and J. Al-Khalili in the Bibliography.

It should be mentioned that besides the Copenhagen interpretation and the Many Worlds theory, there is the De Broglie-Bohm Theory and a string of other interpretations of Quantum Mechanics. Welcome to your vastly mysterious world.

INFINITE SPACE

We are all familiar with one-, two-, and three-dimensional spaces. Any point in a one-dimensional space lies along the real line and is described by its value, a real number, say x. A point in two-dimensional space (the plane) is described by two numbers written as (x, y). A point in three-dimensional space requires coordinates (x, y, z) as in Fig. 7.4. Discussions of spacetime requires four dimensions, (t, x, y, z), where t is time.

Drawing an arrow from the origin to the point (in the direction of the point), also defines a vector in any of the spaces. Vectors in the Euclidean and complex planes were discussed in Chapter 3.

It is useful to note that when multiplying a vector $\mathbf{v} = (x, y)$ by any real *scalar*, say 2, we obtain the vector, $2\mathbf{v} = (2x, 2y)$, and its length will be twice the length of $\mathbf{v}$. That is, the vector is scaled up by the number 2[19]. Try it for yourself. Multiplying by a negative scalar will just point the vector in the opposite direction with a similar scaling effect on its length. So, if we took *all* the real scalar multiples of a

---

[19] For simplicity, we will only consider scalars belonging to the set of real numbers $\mathbb{R}$ or in some cases the complex numbers $\mathbb{C}$, but in general more abstract types of scalars are also allowed.

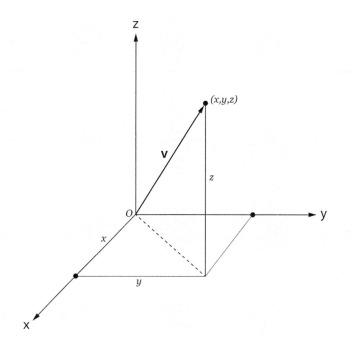

Figure 7.4: A vector **v** in three-dimensional space with coordinates (x, y, z). (Illustration courtesy of Katy Metcalf.)

vector $\mathbf{v} = (x, y)$, we would have an infinite line (ray) through **v**. It is a similar case in three or more dimensions. This should seem pretty obvious.

Now, what about throwing caution to the wind and taking a vector with countably infinite many (real) coordinates? Such a vector will now look like,

$$\mathbf{v} = \left(x_1, x_2, x_3, \ldots, x_n, \ldots \right).$$

Just as in ordinary Euclidean space, we can add two such vectors together by just adding their respective coordinates. We could also multiply such a vector by a (real) scalar number by simply multiplying each coordinate by the number.

So, this new infinite dimensional space will appear very similar to our ordinary three-dimensional space, but with more coordinates. There is one difference however, and that is determining the length (*norm*) of a vector in this new space.

In two dimensions, the vector's length is given by the Pythagorean theorem, i.e. the length of vector $\mathbf{v} = (x, y)$ is

$$\|\mathbf{v}\| = \sqrt{x^2 + y^2},$$

where the two bars on each side of the vector v are just a notation to indicate the vector's length. In three-dimensional space, with another application of Pythagoras (see Fig. 7.4), the length of vector $\mathbf{v} = (x, y, z)$ is likewise

$$\|\mathbf{v}\| = \sqrt{x^2 + y^2 + z^2}.$$

In our infinite dimensional space, the length of vector $\mathbf{v}$ by analogy with Pythagoras would be given by[20]

$$\|\mathbf{v}\| = \sqrt{x_1^2 + x_2^2 + x_3^2 + \ldots + x_n^2 + \ldots} \; .$$

But here we have an infinite sum, which we know from previous considerations is not always finite. Therefore, if we want our space of these vectors to be mathematically respectable, the length of all its vectors should be finite; that is, $\|\mathbf{v}\| < \infty$.[21]

What we have just created is an infinite dimensional space known as ℓ^2. It is called a *Hilbert space*, and is named after its creator, German mathematician David Hilbert (1862−1943). Hilbert is one of the giants of Mathematics, who made many great contributions to the field. At the International Congress of Mathematicians in Paris in 1900, and in the ensuing period, Hilbert announced a list of 23 problems that he felt should be tackled in the new century. Several, including the famous *Riemann Hypothesis* discussed in Chapter 3, remain unsolved.

The space ℓ^2 is just one example of a Hilbert space. The essential properties of such a space are just the right setting in which physicists can do Quantum Mechanics. This was laid out in John von Neumann's masterpiece, *Mathematical Foundations of Quantum Mechanics*, originally published in German in 1932, with an English translation published by Princeton University Press in 1955.

Rather curiously, in a letter to American mathematician George David Birkhoff dated 1935, von Neumann wrote, "I would like to make a confession which may seem immoral: I do not believe in Hilbert space anymore." However, most physicists − and the author − still do.

The general definition of a Hilbert space is not unlike the situation in one-, two-, and three-dimensional Euclidean space. In fact, these three Euclidean spaces *are* Hilbert spaces. In general, one has a collection of objects called vectors that can be added (or subtracted) together to give another vector in the space.

The vectors can also be multiplied by a scalar (say, a real or complex number) to produce another vector in the space. The space allows for a particular kind of

---

[20] We can analogously consider infinite sequences of complex numbers, but then our formula for the length of vector $\mathbf{v}$ will be:

$$\|\mathbf{v}\| = \sqrt{|c_1|^2 + |c_2|^2 + |c_3|^2 + \ldots + |c_n|^2 + \ldots}$$

[21] The correct jargon here is that the coordinates of vector $\mathbf{v} = (x_1, x_2, x_3, \ldots, x_n, \ldots)$ should be 'square summable', since we are summing the squares of all the coordinates of v.

multiplication of two vectors and this multiplication process obeys a few nice properties, with the multiplication returning a scalar value. This multiplication of two vectors allows for the definition of the *'length' (norm)* of a vector, in fact giving the value of the square of its length[22].

This is in complete accordance with the definition of length in our three Euclidean spaces. A final property of a Hilbert space is that it possesses a property called *completeness,* meaning that you will never stray outside the space by taking limits of sequences of vectors inside the space. Again, that is a natural property in order to make your space respectable.

For a physicist doing Quantum Mechanics, a Hilbert space in its mathematical form, as an abstract collection of vectors with a few elementary properties, is as real to them as the air they breathe from three-dimensional space.

Indeed, by employing the bizarre universe of the Quantum Mechanics inhabiting Hilbert spaces, we can produce something equally strange that is slowly becoming very much a part of the real world; namely, quantum computers.

QUBITS

Computers are like humans – they do everything except think...
John von Neumann

As was mentioned in Chapter 4, the basis of any computer are the logic gates comprised of switches, say in the form of transistors, that are either in a voltage state ON, representing the number 1, or state OFF, representing 0.

As an example, simple arithmetic in a computer ends up converting numbers to base 2, so that if we add the numbers 5 + 2, then the computer basically sees the number 5 as 1 0 1, and 2 as 0 1 0, with the appropriate switches ON and OFF[23]. Coupling this with a few base 2 addition rules, again in terms of ON and OFF switches, will give the sum: 111 = 7. A desktop computer can do this just fine, but computations with bytes, composed of bits, is basically done sequentially one at a time.

Let us now enter the strange world of Quantum Mechanics once again. The basic idea is similar to the 0 and 1 bits of a binary computer. We need something

---

[22] In a Euclidean space, this is the *dot (inner) product* of the two vectors. That is if: $\mathbf{u} = (x_1, x_2)$ and $\mathbf{v} = (y_1, y_2)$, then $(\mathbf{u} \cdot \mathbf{v}) = x_1 \, y_1 + x_2 \, y_2$. Thus, $(\mathbf{u} \cdot \mathbf{u}) = x_1^2 + x_2^2 = \|\mathbf{u}\|^2$. Likewise, the inner product induces the norm in a Hilbert space.

[23] There are normally zeros preceding both binary representations but these have been suppressed here. For a *byte* consisting of eight bits, 5 = 00000101, 2 = 00000010, and adding them together gives: 7 = 00000111.

at the quantum level that manifests in two states, like the *spin* of an electron which comes as either 'spin up' or 'spin down'[24]. These two quantum states comprise a *qubit*, designated by, [0] and [1], respectively, and may be combined into what is called a *superposition* of states:

$$\psi = c_1 [0] + c_2 [1]. \tag{27}$$

Here, c_1 and c_2 are *complex* numbers, such that $|c_1|^2$ represents the probability of the electron being in state [0], and $|c_2|^2$ represents the probability of the electron being in state [1]. This is another example of where complex numbers are vital in dealing with problems of the real world.

As we are dealing with probabilities, it is required that: $|c_1|^2 + |c_2|^2 = 1$. Under this proviso, taking all possible complex values c_1 and c_2, we have just created a two-dimensional (complex) Hilbert space. All the vectors, ψ, in this space are represented by the infinitely many different allowable values of c_1 and c_2 in eq. (27).

These vector states can essentially be geometrically represented by what is known as the *Bloch sphere*, as in Fig. 7.5 (left)[25]. When a measurement of the qubit is made, its state will be determined to be either [0] or [1] (due to the collapse of the wave function), and whose value effectively depends on the 'latitude' of the state vector's position on the Bloch sphere.

A quantum computer with 300 qubits can in some sense exist in 2^{300} states *at the same time*. This is more than the number of atoms in the observable Universe[26]. Quantum computers are still in the research and development stage and are suited mainly to particular types of problems (Fig. 7.5 (R)). According to computer scientist Scott Aaronson, the biggest advantages are for "(1) simulating Quantum Physics and Chemistry themselves, which has tremendous potential applications to medicine, materials science, and elsewhere, and (2) breaking the public-key cryptography that's currently used to secure the Internet."

We have just had a peek into the rather mysterious future of computing. Furthermore, quantum computer scientist Seth Lloyd proposed in his book,

---

[24] The notion of spin is derived from the angular momentum of a rotating object, but at the quantum level for a charged particle (like an electron), spin relates to the magnetic field it generates *as if it were* a spinning sphere of charge.

[25] On the Bloch sphere, for θ and ϕ as depicted in Fig. 7.5 (L), there is the geometric representation for the qubit state ψ as:

$$\psi = \sin\frac{\theta}{2}[0] + e^{i\phi}\cos\frac{\theta}{2}[1].$$

The sphere is named after Swiss-American physicist Felix Bloch (1905–1983). Note our familiar friend, the complex exponential $e^{i\phi}$ from Chapter 3.

[26] Note that $2^{300} \approx 10^{90}$, and as mentioned in Chapter 2, the number of atoms in the observable Universe is around 10^{80}.

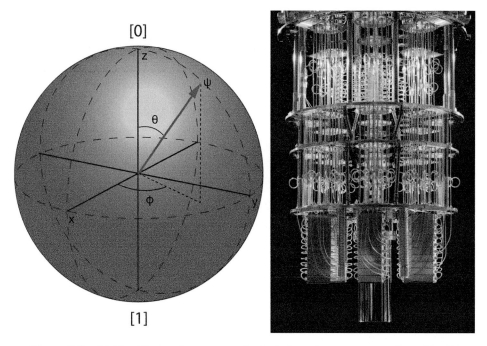

[0]

[1]

Figure 7.5: (L) The Bloch sphere geometric depiction of a state vector of a qubit. (R) Part of the cooling system and cabling of the rather beautiful IBM Q quantum computer. The chip is located in the canister at the bottom, which must be kept near absolute zero. (Images courtesy of (L) Katy Metcalf, (R) IBM Research *Flickr* CC BY-ND 2.0.)

Programming the Universe, that the entire Universe is a quantum computer that "computes its own behavior." This idea is similar to the digital Physics of Edward Fredkin mentioned in Chapter 5, in the context of the Universe being a cellular automaton.

Related to this latter idea is the notion that in order to simulate behavior in the microscopic quantum world effectively, it will take a quantum computer to do so. In the words of theoretical physicist John Preskill, "…might we already have persuasive evidence that Nature performs tasks going beyond what can be simulated efficiently by classical computers?" This is the quantum computing version of the expression, "It takes a thief to catch a thief."

Preskill proposed the term 'quantum supremacy' in 2012 to denote the point in time when a quantum computer will be able to perform a computing task that would not be possible on a classical computer *in a reasonable period of time*, irrespective of whether or not the computation was in any way useful.

The achievement of this goal of quantum supremacy has been claimed by an AI team at Google, published in an article in *Nature* (October 23, 2019), who used their

53-qubit quantum computer chip called Sycamore. They have performed a calculation in roughly 200 seconds that they argued would have taken the most powerful supercomputer around 10,000 years to perform. It must be added, however, that the problem selected was designed very specifically to favor a quantum computer and is known to be difficult for a classical computer. The authors claimed[27]:

> "Quantum processors based on superconducting qubits can now perform computations in a Hilbert space of dimension $2^{53} \approx 9 \times 10^{15}$, beyond the reach of the fastest classical supercomputers available today. To our knowledge, this experiment marks the first computation that can be performed only on a quantum processor."

On the other hand, in an IBM research blog[28], three of their scientists asserted that the problem Google solved could have been done in 2.5 days on one of its classical supercomputers, which is rather quicker than Google's estimate of 10,000 years. Therefore, "… [Google's assertion] should not be viewed as proof that quantum computers are 'supreme' over classical computers." Needless to say, IBM and Google are fierce rivals in the field of quantum computing.

Part of the problem is the very notion of quantum supremacy and its interpretation. Moreover, whether the computer time is 10,000 years or 2.5 days is not really the point. The IBM authors go on to state that, "quantum computers will never reign 'supreme' over classical computers, but will rather work in concert with them, since each have their unique strengths." Indeed, some problems are more amenable to classical computers and some are more suitable for a quantum computer, though the line between the two types of problems is often a gray area.

What the Google result does demonstrate is control over 53 qubits that are inherently unstable and were held in a quantum superposition state long enough to do a complex calculation. So, achieving this sort of result is an impressive technical achievement and a scientific milestone pointing the way to the future.

For those interested, there are now facilities for anyone from the general public who wishes to learn how to create a program on a quantum computer and to execute it[29]. A comprehensive account of quantum computing, together with much related subject matter, can be found in the fine monograph by Scott Aaronson in the Bibliography.

---

[27] F. Arute *et al.*, Quantum supremacy using a programmable superconducting processor, *Nature*, 574, 505-510, 2019.

[28] E. Pednault, J. Gunnels, J. Gambetta, On Quantum Supremacy, Oct. 21, 2019, IBM Research blog. This announcement pre-dates the Google publication in *Nature*, as the Google announcement was leaked publicly prior to publication.

[29] D-Wave Leap: https://www.dwavesys.com/take-leap
IBM Q Experience: https://www.research.ibm.com/ibm-q/Microsoft Quantum Development Kit: https://www.microsoft.com/en-us/quantum/development-kit

IT IS ALL RELATIVE, ALBERT

When you sit with a nice girl for two hours, you think it's only a minute. But when you sit on a hot stove for a minute, you think it's two hours. That's relativity...
Albert Einstein

When the author was a small boy and there was a bolt of lightning, he was told to count the number of seconds between the flash of lightning and the clap of thunder that followed later. Then, dividing by 5 gave the approximate number of miles away the lightning strike was. When he was older, it was apparent to the author that the flash reached his eyes almost instantaneously, but with the sound traveling at roughly 1,000 feet per second (more accurately 1,150 ft/sec), it meant that in 5 seconds the sound would travel about 5,000 ft, or nearly a mile. So, if you count 15 seconds between the lightning and the thunder, you can rest assured that you are about three miles away from the action.

In this instance, the sound from the thunder is rather lumberingly slow compared to the lightning, which is blindingly fast at 300,000 km/sec. In fact, this is the maximum speed limit in our Universe for moving objects, and all electromagnetic radiation, from radio waves to light waves, to the waves of microwave ovens, travels at this same speed. Nevertheless, the speed is finite and so must have some consequences, as the finite speed of sound does in the example above.

However, light has one curious property that we do not experience in everyday life. If you are on a bus traveling, say, at 50 km/hr and you get up and walk to the front of the bus with a speed of 5 km/hr, then your speed relative to a person standing on the roadway would be the sum of the two speeds, namely 55 km/hr. This is intuitively easy to grasp. Alternatively, if you walk towards the back of the bus at 5 km/hr, then your speed would be 45 km/hr relative to the person on the roadway.

But this is not the case with light. If you were on a very fast bus traveling, say, at 50,000 km/sec (just for the sake of argument), and took out a flashlight and shone the beam ahead, the light's speed as measured by a person on the ground would still be 300,000 km/sec. Similarly, if you pointed the flashlight towards the back of the bus, the light's velocity measured by someone on the ground would be the same. Not only that, if you conducted an experiment to measure the velocity of light while traveling on this magical bus, you would find that the speed of light was again 300,000 km/sec.

Here we have one of the two main postulates upon which the Special Theory of Relativity is built:

The speed of light is constant when measured by any observer in the Universe and is independent of the motion of the light source.

This has been experimentally verified, most famously by the Michelson-Morley experiment in 1887 (Michelson-Morley set out to prove something entirely different – namely the existence of an *aether* medium that carried light waves), and by more modern experiments using lasers, masers, and other sophisticated devices. Moreover, not only is the speed of light constant, but at the Conference Generale des Poids et Mesures in 1983, it was *defined* to be 299,792,458 meters/second. So, the speed of light cannot be any more constant than that.

The second postulate of the Special Theory of Relativity is the following:

The laws of Physics are the same for all observers in uniform motion with respect to one another.

This one is slightly more arcane, but 'uniform motion' just means moving at a constant velocity in a straight line, like two passing trains. What the postulate says is that any Physics experiments performed aboard one train will give the same results when performed on the other train. In other words, the laws of Physics will be the same on both trains.

In what follows, we are going to discuss some aspects of Einstein's Theory of Relativity. This theory, first enunciated by unknown Swiss patent office clerk, technical expert third class Albert Einstein in 1905 (Special Theory of Relativity without gravity), and then again in 1915/1916 (General Theory of Relativity including gravity), has revolutionized in a very profound way how we think about time and space, and the Universe at large.

In the past, the theory has been mainly of interest to those who are serious students of Physics, and there was little in the Theory of Relativity that had any real impact on the everyday lives of mortal beings. This was because the effects of Relativity are only manifested under extreme conditions which are not experienced in everyday life, or are so imperceptible as to go unnoticed except in sophisticated laboratories.

However, all that has changed with the advent of the Global Positioning System (GPS) that is found in many cell phones and on the dashboards of many cars. So, this arcane theory has impinged directly on the lives of many of us and GPS will be explained in due course. But first, let us consider a specific example concerning time.

In his 1905 Special Theory of Relativity, Einstein was the first to draw attention to the issue of simultaneity; that is, what do we mean by two events occurring at *exactly* the same time.

Let us consider the following thought experiment that is based on an account of the actual 1918 signing of the Armistice of World War I. This was to take place in a railroad carriage in a French forest at 05:00, with hostilities to cease at 11:00 that same day: the 11th hour of the 11th day of the 11th month.

The author saw this example featured in a documentary on Relativity many years ago and has enjoyed it ever since. Let us now suppose that the Armistice signing was to take place at the appointed hour, with both the Allies, headed by

Marshall Ferdinand Foch, and the German side, headed by Matthias Erzberger, signing the Armistice agreement at *exactly* the same time.

Taking some hypothetical liberties, the Allies are seated at one end of the carriage and the Germans are seated at the opposite end. Let us also suppose that each party is signing a separate copy of the Armistice agreement and let us now introduce some observers, one sitting next to the Allies and one next to the Germans, in order to verify that both parties sign at *precisely* the same time.

A clock strikes the appointed hour and we must now suspend any concerns about reaction times and the like and assume that both signing parties are able to put pen to paper instantaneously with the strike of the clock. Of course, this is not a true-life situation, but is just meant to illustrate an important scientific notion, and one we simply take for granted in our everyday lives.

Now remember that light, although traveling at phenomenal speed, still takes a bit of time to travel from one place to another. In fact, the sunlight currently illuminating the author at his computer left the Sun over eight minutes ago, while at night, the light hitting his eye from the nearest star left its host over four years ago[30]. The light beams from our beautiful neighboring Andromeda galaxy that are just reaching us now left their mother galaxy some 2.5 million years ago. Thus, on a cosmic scale, the finite speed of light has a lot of ramifications. But this is rarely the case in our daily life, and certainly not in a carriage on a railroad siding in a French forest. Be that as it may, let us not be deterred in our examination of the Armistice signing.

Precisely at the stroke of 05:00, the two parties begin to sign the Armistice treaty. After the signing takes place, the observer on the Allied side claims that the signing was not fair as the Allies started signing the treaty first. Why would he say that? Well, sitting next to the Allies, the tip of the pen used by Marshall Foch is closer to him than the tip of the pen of Erzberger. Thus, the light from Erzberger's pen would have taken a little bit longer to reach him and so he would have seen Foch signing first. But no, says the observer sitting with the German side. He saw the Germans signing first, as the tip of Erzberger's pen was closer to him than the tip of Foch's pen, and so he saw Erzberger signing first. Here we have two different observers, and each saying the complete opposite of the other.

Really, in all fairness, let us introduce a third observer who is exactly half way between the Allied and German sides. He is looking in the direction of the German side, but also holds up a mirror so that he can also see the Allied side behind him. At the stroke of the clock, he sees both the Allies and Germans signing the Armistice Treaty at precisely the same time. This is because he is equally distant between both of them and the light from the pens of Foch and Erzberger take exactly the same time to reach him.

---

[30] The nearest visible star is Alpha Centauri which is actually a close binary system consisting of two stars at a distance of 4.35 light-years. A third star, Proxima Centauri, orbits the other two and is somewhat closer to the Earth but is not visible to the naked eye.

So, did the Allies and Germans in our little dramatization sign at the same time? Well, it is all relative now is it not? Why is it all relative? Because light, just like sound, has a finite speed and this fact has consequences, although we are normally not even aware of them. However, one of the consequences (the distortion of time), is utilized by many of us all the time, as we will see in due course.

The situation of simultaneity becomes even more interesting if the railway carriage is actually in motion, and the following example comes directly from Einstein's Special Theory of Relativity (1905).

Let us suppose that a train has a front and back, labeled A and B respectively, and is moving along a straight track at a high constant velocity. There is one observer in the middle of the train and one on a platform as the train speeds past. Just as the train passenger and platform observer are directly opposite one another, let us suppose that two bolts of lightning strike the points A and B on the train simultaneously, according to the platform observer, who at that very instant happens to be equidistant from both ends of the train (Fig. 7.6). What does the observer on the train witness?

Figure 7.6: Lightning strikes the front and back of a fast-moving train 'simultaneously' according to the person standing on the platform, who is half-way from each end of the train at the time. But a person sitting at the middle of the train will witness the strike at the front of the train first and so claim that it occurred before the strike at the rear. (Illustration courtesy of Katy Metcalf.)

Since the train is moving forward, the light from the lightning strike at the head of the train A will reach the mid-train observer first, as the observer is moving towards that light and thus the light has had a shorter distance to travel than the light emitted at the back of the train, which has had to catch up with the mid-train observer. Therefore, the mid-train observer will see the lightning strike the front of the train *first* and will conclude that the strike at A preceded the strike at B.

However, the platform observer will say that the lightning strikes at A and B were simultaneous. Which observer is correct? Well, they are both correct, from their respective reference frames. Again, it is all relative.

Another consequence of the Special Theory of Relativity is a very simple equation, perhaps the most widely known equation on Earth.

THAT EQUATION

Equations are more important to me, because politics is for the present,
but an equation is something for eternity ...
Albert Einstein

Figure 7.7: Yes, *that* equation.

One of the consequences of Einstein's great work on Special Relativity is the famous equation

$$E = mc^2,$$

which represents one of the most profound expressions about our Universe in all of Science (Fig. 7.7). It appeared in 1905 in the article, *Ist die Trägheit eines Körpers von seinem Energieinhalt abhängig?* (Does the inertia of a body depend upon its energy-content?). Although credit must also be given to predecessors Olinto De Pretto, Friedrich Hasenöhrl, and Henri Poincaré, who had earlier discovered the same equation in different settings. In Einstein's own words some years later[31]:

"It followed from the Special Theory of Relativity that mass and energy are both but different manifestations of the same thing – a somewhat unfamiliar conception for the average mind. Furthermore, the equation *E* is equal to mc-squared, in which energy is put equal to mass, multiplied by the square of the velocity of light, showed that very small amounts of mass may be

---

[31] From the 1948 documentary, *Atomic Physics*, United World Films for the J. Arthur Rank Organization, Ltd, https://www.youtube.com/watch?v=JzRvCkn8KL8.

converted into a very large amount of energy and vice versa. The mass and energy were in fact equivalent, according to the formula mentioned above. This was demonstrated by Cockcroft and Walton in 1932, experimentally."

If a single gram (0.035 oz) of mass were converted entirely to energy, it would be equivalent to 21,000 tons of TNT, which is the energy yield of the atomic bomb dropped on Nagasaki in Japan during World War II. The important component of the preceding equation is the c^2, which is the velocity of light multiplied by itself, an exceedingly enormous number.

Approximately four million tons of the Sun's mass is converted to energy every second, thus permitting life to exist on Earth. Recent experiments have also verified the exactness of the Einstein equation to four-tenths of one part in a million[32].

On a smaller scale, $E = mc^2$ is behind the production of nuclear power. Controlled nuclear reactions in nuclear power plants release energy that generates heat. The heat is used to turn water into pressurized steam that turns a turbine which, combined with a generator, produces electricity. All from that one little equation.

We next turn our attention to the notion of time itself, as both Special and General Relativity affect time in their own unique way. Knowing the precise details of those effects allows us to navigate via GPS and get from A to B.

WHAT TIME IS IT ANYWAY?

Time is what keeps everything from happening at once…
Science fiction writer Ray Cummings

Excuse me, do you have the time? The time is now, child…
Nepalese woman answering query of a young boy[33].

Most of us take time for granted. It is something we get from watches and clocks, and we all have a past and hopefully a future, but we say that we live in the 'now'. What time actually *is* mainly concerns scientists and philosophers, and no one really knows the answer to that question. Time might not even run smoothly and continuously, but could be broken up into discrete units that accumulate one upon the other. But there is one thing we probably would all agree upon, which is that time cannot speed up or slow down. It might *seem* like time slows down when we are toiling at a job we dislike, or speed up when out on a wonderful date, but we know that it is not time that is altering, merely our perception of it.

---

[32] S. Rainville, *et al.*, A direct test of E = mc², *Nature*, Dec. 22, 2005.

[33] From a short *National Geographic* film:
https://www.youtube.com/watch?v=R3VMW6fxK6Y

Thus, it would be very surprising to find out that time *can* actually slow down or speed up, and that these 'time dilations' have actually been measured in many different experiments. These very strange phenomena were first delineated in Einstein's Special Theory of Relativity, which showed that moving clocks will appear to tick slower when compared to a stationary counterpart. Moreover, Einstein (1907) also demonstrated how gravity slows down time. It is fair to say that time, as it is commonly understood, has never been the same since Einstein. These notions of time dilation are one of the key ingredients in the movie *Interstellar.*

Thus, the pop song by the singer Cher, "*If I Could Turn Back Time*" is partially realizable, in that it is possible to slow down time, but one either has to be travelling exceeding fast or go and live on Jupiter with its intense gravity. However, the latter will have undesirable effects on one's physiology.

So, besides keeping us firmly planted on Earth, let us first consider what is going on with gravity?

MATTERS OF GRAVITY

> *Gravitation is not responsible for people falling in love…*
> Albert Einstein

Everyone on Earth has plenty of experience with gravity – even bacteria – and, curiously enough, time itself (this will be explained later). We all experience the effects of gravity that manifests itself in our own weight, or if we attempt to jump off the planet. But for thousands of years, no one gave much thought to the notion of gravity, or why things fell down and not up.

That is until Isaac Newton was forced to return home from Cambridge University due to the Great Plague. During this spell (1665–1667) at his childhood home at Woolsthorpe Manor, there occurred the now legendary incident of the falling apple. Many years later, the details of the event were verbally related by Newton to biographer William Stukeley:

"After dinner, the weather being warm, we went into the garden, and drank tea under the shade of some apple trees, only he, and myself. Amidst other discourse, he told me, he was just in the same situation, as when formerly, the notion of gravitation came into his mind. 'Why should that apple always descend perpendicularly to the ground', thought he to himself, occasioned by the fall of an apple, as he sat in a contemplative mood. 'Why should it not go sideways, or upwards but constantly to the earth's center? Assuredly, the reason is, that the earth draws it. There must be a drawing power in matter, and the sum of the drawing power in the matter of the earth must be in the earth's center, not in any side of the earth. Therefore, does this apple fall perpendicularly, or toward the center. If matter thus draws matter; it must be in proportion of its quantity. Therefore, the apple draws the earth, as well as the earth draws the apple'[34]."

On the other hand, there is no mention that the apple actually fell on Newton's head. It was Newton's great genius that allowed him to extend this 'drawing power' to the entire Universe.

Newton's Law of Universal Gravitation[35] says that any two objects will attract one another with a force that is proportional to the product of the masses, and inversely proportional to the square of the distance between their centers of mass[36]. It can be expressed by the simple formula

$$F = G\frac{m_1 m_2}{r^2} \tag{28}$$

where m_1, m_2 are the values of the two masses, r is the distance between them, and G is the universal constant of gravitation[37].

This constant G has an exceedingly small value and that is why the force of gravity is not noticed between everyday objects. Even if you stand alongside a mountain, you will not notice being pulled towards it.

The reason for such 'inverse square' behavior that is found in various physical phenomena, such as gravity, the electrostatic force, or electromagnetic radiation, is that the particular quantity (force, energy, etc.) is being radiated from its source in all directions into three-dimensional space. As an example, let us observe the energy intensity on the surface of an expanding sphere for an energy point source at the center of the sphere, as in Fig. 7.8.

The formula for the surface area S of a sphere of radius r is given by

$$S = 4\pi r^2.$$

---

[34] From William Stukeley, *Memoirs of Sir Isaac Newton's Life*, 1752. Some punctuation and spelling have been modernized.

[35] Published in 1687 in his immortal work, *Philosophiæ Naturalis Principia Mathematica*.

[36] For large spherical bodies like planets and stars, it is justifiable to assume that all their mass is concentrated at their respective centers.

[37] $G = 6.674 \times 10^{-11} \ m^3 \cdot kg^{-1} \cdot sec^{-2}$.
A similar formulation (known as *Coulomb's Law*) to eq. (28) describes the *electrostatic force*, F, between two (stationary) point charges,

$$F = k\frac{q_1 q_2}{r^2},$$

where k is Coulomb's constant, q_1 and q_2 are the respective magnitudes of the two charges (which can be + or −), and r is the distance between them. If the signs are opposite, the force is attractive, and if the signs are the same, then the force is repulsive.

The *intensity* of the energy E emitted by a point source at the sphere's center is the *incident energy per unit of area on the sphere*; that is, we must divide the energy E by the area of the sphere to get the intensity

$$Intensity = \frac{E}{4\pi r^2}. \tag{29}$$

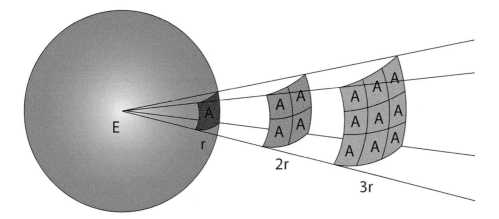

Figure 7.8: Doubling the distance from the source diminishes the energy by a factor of four and tripling the distance diminishes the energy by a factor of nine and so forth, according to the inverse square of the distance. (Illustration courtesy of Katy Metcalf.)

Therefore, as the distance r from the source increases, the energy per unit area (intensity) on the sphere is being diluted according to eq. (29) by the inverse of the square of the distance r. As a consequence, doubling the distance from the source reduces the energy's intensity by a factor of four; tripling the distance reduces the intensity by a factor of nine, and so on, according to the inverse square of the distance (Fig. 7.8).

Gravity was still an unexplained mysterious force that acted instantaneously between any two objects, like the Sun and Earth, but at least there was a formula to describe this attractive force, and it worked very well in describing the motions of the planets – up to a point.

Unfortunately, it could not explain certain discrepancies in the orbit of Mercury, namely that the elliptical orbit that Mercury moves in about the Sun was seen to revolve gradually (known as *precession*), as depicted in Fig. 7.9.

This rotation of Mercury's orbit was due to the effects of the other planets (530 arc seconds per century), but some 43 arc seconds per century could not be explained by Newton's Law alone.

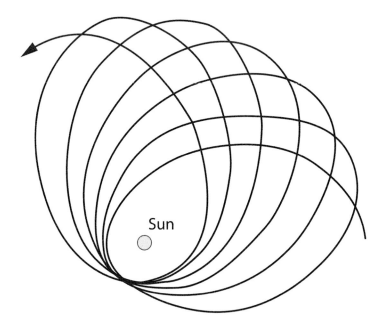

Figure 7.9: (Exaggerated) illustration of the precession of the orbit of the planet Mercury about the Sun. (Illustration courtesy of Katy Metcalf.)

Actually, according to Einstein's General Theory of Relativity which he delivered to the world in 1915, gravity it is not really a force as Newton conceived it, but is the result of the 'bending of spacetime', which means that both the space and time around a massive object are actually warped by the object, analogous to the way a bowling ball would distort the surface of a trampoline[38]. Moreover, relativity could completely account for the discrepancy in the precession of the orbit of Mercury.

Interestingly, this same phenomenon of orbital precession has been observed in a star, famously known as S2, orbiting the black hole Sagittarius A* at the center of the Milky Way, again in complete accordance with General Relativity.

---

[38] Spacetime is the mathematical fusion of three-dimensional space with one-dimensional time, creating a four-dimensional continuum. Any point in spacetime will have three spatial dimensions and one temporal dimension. For Minkowski Spacetime, see Appendix II.

Einstein's profound revelations about the workings of the Universe came about in the form of a set of ten mathematical equations of General Relativity, which can be written in a condensed form

$$R_{\mu\nu} - \frac{1}{2}g_{\mu\nu}R + g_{\mu\nu}\Lambda = \frac{8\pi G}{c^4}T_{\mu\nu}. \tag{30}$$

Without going into the arcane details, note that there are some old friends here, like π, which has wormed its way into the most fundamental description of the Universe. Also familiar is the velocity of light given by the letter, c, and the capital G is the traditional gravitational constant. The Greek letter Λ is the *cosmological constant* that represents an expansionary force due to dark energy, which will be discussed later in the text.

Originally, Einstein put in the cosmological constant because it was thought at the time that the Universe was static; neither expanding nor contracting. The cosmological constant, being an expansionary force, put the brakes on gravity due to the accumulated matter in the Universe, which would cause it to contract. However, a series of subsequent developments showed that the Universe was actually expanding.

The first such evidence came from Dutch mathematical physicist Willem de Sitter (1832–1934), whom we encountered in Chapter 1 with the anti-de Sitter model. De Sitter's 1917 solution to Einstein's equations of General Relativity made the assumption of a Universe being devoid of ordinary matter, but whose dynamics were dominated by the (positive) cosmological constant Λ.[39] This led to the de Sitter model of the Universe that was expanding.

Filling the Universe with ordinary matter, coupled with a cosmological constant, in 1922 Russian mathematician Alexander Friedmann showed that, from the Generality Relativity equations, one could still derive a dynamic expanding Universe (see Footnote 19 in Chapter 6). However, Friedmann's work received little attention at the time. Then, in 1927, Belgian priest and mathematician Georges Lemaître (1894–1966) also derived an expanding Universe from Einstein's equations. Furthermore, Lemaître discovered the linear relationship between the velocity of expansion of the Universe and distance that is now known as Hubble's Law (eq. 22). Lemaître even made an estimate of the constant of proportionality, which is now called Hubble's constant H_0, as we have seen earlier.

Unfortunately, for Lemaître, his work was published in French, in relative obscurity in the *Annales de la Société scientifique de Bruxelles*, and was only translated into English in 1931. The translated version of the paper also left out two key paragraphs concerning Lemaitre's discovery, which noted that the velocity of expansion of the Universe is proportional to distance, something he also

---

[39] By contrast, the anti-de Sitter model of the Universe has a negative cosmological constant.

based on Vesto Slipher's redshift data, as had Hubble. This led to decades of speculation that some skullduggery had been involved during the translation process, in order to give all the credit to Hubble.

Only in 2011 did investigations by scientist Mario Livio finally resolve the truth of the matter[40]. Lemaître *himself* had translated the original paper from the French, and it was he who had left out the crucial sections dealing with the constant of proportionality determining the rate of expansion of the Universe as a function of distance. It seems that, by 1931, Lemaître was not overly concerned that Hubble had already been given prior credit for discovering arguably one of the most important constants governing the workings of the Universe. The upshot of all of this is that by 1929, it was American astronomer Edwin Hubble (1889−1953), who had garnered most of the credit for demonstrating the expansion of the Universe.

Lurking in the background, there will often be considerable drama surrounding many scientific discoveries. In general, it is complicated, as we have already seen with the Mandelbrot set and the discovery of Neptune. Indeed, the current model of the Universe accepted by most cosmologists is called the Friedmann-Lemaître-Robertson-Walker (FLRW) model.

So, some familiar ground is to be found in eq. (30), but the rest is very abstruse and will not be pursued further. Basically, the left-hand side describes the way in which spacetime is curved by the mass and energy as described on the right-hand side. Most people would simply find these equations filled with squiggles, and yet the fascinating things is, these very squiggles can predict how the Universe works on the largest scales, from planets and moons, to galaxies and starlight. This is one of the most profound descriptions of the world available to Science. According to the brilliant American physicist John Wheeler: "Space tells matter how to move. Matter tells space how to curve." Though technically one should replace 'space' with 'spacetime' in the quote, we need not quibble with Wheeler.

The first experimental verification of this revolutionary notion came in 1919, when English astrophysicist Arthur Eddington mounted an expedition to South Africa to observe a total eclipse of the Sun on May 29 from the island of Principe. At the same time, another expedition party led by Sir Frank Watson Dyson, the Astronomer Royal, took a telescope to Sobral in northern Brazil.

Both telescopes were meant to measure accurately the position in the sky of several stars near the periphery of the disk of the Sun during the eclipse, which normally could not be seen. According to Einstein, the space around the Sun is warped so that starlight which passes near the Sun's disk will follow a curved trajectory to us on the ground. Thus, the starlight will appear displaced from its normal position when the Sun does not intervene (Fig. 7.10).

The predicted light displacement was 1.75 arc seconds, a very infinitesimal but measurable quantity at the time. Newton's theory of gravity predicted a value of

---

[40] M. Livio, Mystery of the missing text solved, *Nature*, Vol. 479, pp. 171−173, 2011.

half this amount. The actual verification by Eddington made front page news (Fig. 7.11) and made Einstein a scientific superstar. The innate 'correctness' of these equations was so strongly felt by Einstein that when he was asked what would be his response if he had not been proven right, he replied, "Then I would feel sorry for the dear Lord. The theory is correct anyway."

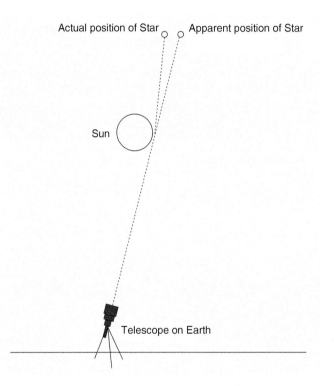

Figure 7.10: Depiction of the bending of starlight by the Sun due to the warping of space in its near vicinity. (Illustration courtesy of Katy Metcalf.)

The actual results obtained by Eddington were controversial at the time. Eddington was accused of bias and of manipulating the data (claimed to be of poor quality), as he was a strong advocate of Einstein's theory. Indeed he was one of the very few people in the world to understand it. When told in 1919 that only three people in the world understood the Theory of Relativity, Eddington is reported to have said, "Who is the third?"

But fudge the data he did not, and in spite of various measurement complications, the data did vindicate General Relativity, as have many eclipses of the Sun ever since. The full story of the 1919 expedition and the measurement issues

Figure 7.11: (L) Front page of the *New York Times*, November 10, 1919. (R) Image from the expedition to verify Einstein's prediction of the bending of light around the Sun. Vertical marks indicate position of stars measured by Eddington. (Images in public domain.)

involved is quite interesting, and like many scientific discoveries, it is not without drama[41].

This bending of space in accordance with the General Theory of Relativity has actually been verified with other types of experiments, including the *frequency shift* of radio waves from the Cassini spacecraft, caused by the Sun, which agreed with the Theory to four decimal places[42]. This notion will be discussed further subsequently.

Another notable consequence of the warping of spacetime is the slowing down of time by gravity (*time dilation*), which we alluded to earlier and which has also been experimentally verified[43]. What happens with time is that a clock near the Earth's surface will tick a fraction slower than a clock a bit higher up due to the difference in gravitational potential. This fact will account for half the story regarding the workings of the GPS system to be discussed in the next section.

---

[41] A fine discussion of this issue can be found in the article by Daniel Kennefick: Not Only Because of Theory: Dyson, Eddington and the Competing Myths of the 1919 Eclipse Expedition, https://arxiv.org/abs/0709.0685

[42] Bertotti, *et al.*, A test of General Relativity using radio links with the Cassini spacecraft, *Nature,* Vol. 425, pp. 374–376, 2003.

[43] This was actually demonstrated mathematically by Einstein in 1907, eight years before the publication of his General Theory of Relativity.

You might think that we are referring to ordinary mechanical clocks and that somehow the apparatus inside is more sluggish on Earth, thus slowing it down. But mechanical clocks based on the rotation of the Earth around the Sun are of little use in Science and instead scientists use what are called 'atomic clocks' (Fig. 7.12). These are often based on the natural oscillations of the cesium (also spelled cae-sium) atom, which 'ticks' at just over nine billion times per second. In fact, 1 second is actually *defined* to be 9,192,631,770 ticks of the cesium-133 atom.

Figure 7.12: JILA laboratory's experimental atomic clock based on strontium atoms held in a lattice of laser light is the world's most precise and stable atomic clock. It should keep 'perfect' time for five billion years assuming the equipment lasts. The image is a composite of many photos taken with long exposure times and other techniques to make the lasers more visible. (Image courtesy of Ye group and Baxley/JILA.)

An extraordinary demonstration of time dilation was revealed in a 2010 experiment by James Chin-Wen Chou and his team at the U.S. National Institute of Standards and Technology. They had two very accurate clocks, in each of which the oscillations of a single aluminum ion – an electrically charged atom – were measured by a laser. The two identical atomic clocks were a mere 33 centimeters (about 13 inches) different in elevation. The *lower* clock experienced an ever so slightly stronger pull of the Earth's gravity than the higher clock, and as a consequence lagged fractionally behind the higher one. In fact, the higher one ran a very discernible 4 parts in 10^{17} faster.

The fact that human beings can even measure this difference is remarkable, and the fact that a relativistic time dilation manifests itself at such a small scale (a 33 cm height difference) gives a fascinating glimpse into the subtle deformation of spacetime right under our very noses.

What this also signifies is that every region of spacetime has its own 'local time' that depends upon the gravitational influences acting upon it. Thus, there is no such thing as 'universal time'. For ordinary purposes, however, there is no need to adjust your watch whenever you climb up a ladder. You can also check the time on your computer with a clock synchronized with an atomic clock at: https://time.is/about

TIME IN MOTION

It would seem that the Theory of Relativity is rather arcane – which it is – and has nothing to do with our ordinary lives. But it does. As it turns out, many of us make daily use of the accuracy of atomic clocks, and that is whenever we use a GPS device. The GPS system consists of a network of 31 satellites (upgraded from the original 24) orbiting the Earth, which carry onboard atomic clocks in order to know the time to a very high degree of accuracy. The satellites are arrayed in space in such a way that there are six satellites within view of any location on Earth at any given time. Each satellite broadcasts a radio signal that contains the information of the time it was sent (a *timestamp*), and the location of the satellite in its orbit.

Your GPS unit contains a basic quartz clock since atomic clocks are rather expensive[44]. This clock does not have the required accuracy to remain synchronized with the atomic clocks, but it is constantly being reset (when turned on) with the atomic clocks on board the satellites. This can be done by your GPS connecting with four or more satellites, and its clock is then able to correct its own inaccuracy, but we omit the fine details.

By measuring the difference between when the time signal was sent, given by the timestamp, and the time of its arrival on Earth, the distance from you to the satellite is computed by your GPS. This is because these radio signals travel at the known speed of light, c, and so just like sound, travel time determines distance.

Furthermore, using the timing and location information from a set of three such satellites and a process known as 'trilateration', your latitude/longitude position on Earth can be determined to within several meters[45]. Add in the information from a fourth satellite, and even your elevation can be determined as well.

---

[44] Much of the GPS technology has now been built into smartphones.

[45] If you are a distance a from satellite A, then you exist somewhere on a sphere centered about A of radius a. Similarly, for distance b from satellite B, and distance c from satellite C. The intersection of these three respective spheres will be two points, but only one will be on the surface of the Earth, which is your latitude and longitude position.

But the clocks in the satellites must be synchronized with clocks on Earth, and to do that requires that we use Einstein's Theory of Relativity. This is because the clock aboard a satellite will be running a little faster than a stationary clock on Earth, as we have seen, due to the ground-based stationary clock experiencing higher gravity than the one aboard a satellite orbiting at an altitude of 20,200 km.

On the other hand, this effect of the satellite clock running faster will be offset somewhat by the fact that the satellite is in orbit with a velocity of some 14,000 km/hr. This means that a satellite's clock should also run a little bit slower than a stationary one. To see why this is so, we need Relativity again.

To understand this latter point, suppose that we create a new type of clock using a beam of light instead of a pendulum swinging back and forth or ions oscillating. The advantage of using such a clock is that we do not have to construct it physically, we can just draw a simple image of it (Fig. 7.13).

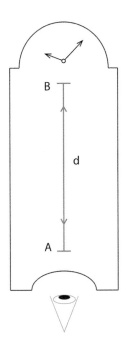

Figure 7.13: A light clock where a 'tick' is the time taken by the light beam to go from A to B and back to A again. (Illustration courtesy of Katy Metcalf.)

To make our clock work, we bounce the beam of light between two mirrors located at positions A and B and record the time the light beam takes to return to A again. Let us call that one 'tick' of the clock, analogous to an interval of time, say, one second. The duration of the tick of course depends on the fixed distance d that the mirrors are from each other, but its actual value is of no importance to us. It is just some interval of time t_0, and we will not alter it in this experiment.

That covers our stationary clock. Now let us put the light clock on a rocket ship, moving at a constant velocity v.

We will measure the duration of a single tick of our light-beam clock once more, where again the distance d apart of the mirrors is the same distance from the light source as in Fig. 7.13.

The only difference in the moving rocket scenario is that the light beam will appear to someone on the ground to take a somewhat longer path, as indicated in Fig. 7.14. One can visualize this by replacing the mirror clock with a person bobbing a yo-yo up and down. To the person holding the yo-yo, it is still just going up and down, but to an observer on the ground, the yo-yo is tracing out a zigzag path as in Fig. 7.14 (top right). The added virtue of this type of 'thought experiment' is that we do not have to build a rocket ship either.

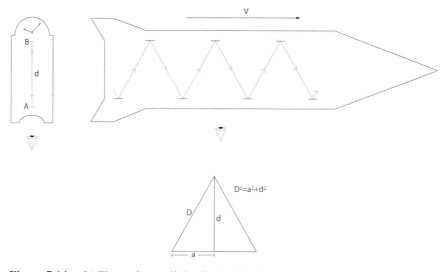

Figure 7.14: (L) The stationary light clock as in Fig 7.13. (R) The same clock aboard a moving rocket ship showing only the path of the light wave. The path traced out by the light beam seen by an observer on the ground will take the zigzag path indicated. (Bottom). Since the length of the hypotenuse D is greater than the side of length d, a 'tick' on board the moving rocket will appear longer to a stationary observer on the ground, and thus rocket-time will appear to run slower than for someone on the ground. (Illustration courtesy of Katy Metcalf.)

Again, let us measure the tick of the moving clock from our ground observer's perspective. For half a tick of the clock, the apparent distance travelled is now D, as in the Fig. 7.14 (bottom). You can see that we are dealing with a right-angled triangle – actually two of them with identical dimensions. Since the length of the hypotenuse D is greater than the side of length d, a 'tick' of the clock aboard the moving rocket will appear to have longer duration for a stationary observer on the ground.

Thanks to Pythagoras, we can compute exactly how time aboard the rocket is related to the time of a stationary clock. This is so because the two light clock distances between each tick are $2D$ and $2d$ respectively, and these are related by the Pythagorean theorem as in Fig. 7.14 (bottom). With a touch of algebra (see Appendix XV for the details), we are able to determine the relationship between the time t aboard a rocket moving at a constant velocity v, compared to the time t_0 computed by a stationary observer on the ground. The all-important formula is called the *time (Lorentz) dilation factor*

$$t = \frac{t_0}{\sqrt{1 - v^2/c^2}}. \tag{31}$$

Here, we see that a formula from antiquity (Pythagoras' theorem for ideal perfect triangles) directly impinges on the deepest notions of 20th century science – Relativity[46]. An inspired use of this formula in a simple thought experiment turns out to reveal something very fundamental about the Universe.

What does eq. (31) tell us about the time duration t in our moving rocket ship? Firstly, note that when $v = 0$, then both clocks tick at the same rate, $t = t_0$.

In general, the denominator in eq. (31) is always less than 1 in a moving setting and that means

$$t > t_0.$$

This is not a result we would naturally expect. We have demonstrated that the time duration of a tick of our clock has lengthened in our moving framework relative to our stationary framework. If each individual tick of the clock takes longer in duration, that means that time itself has slowed down relative to our stationary framework. Time once again is seen not to be an immutable quantity. It is variable, and the amount it slows down depends on how fast the clock aboard the rocket ship is traveling (Fig. 7.15).

For example, if the rocket is traveling at, say, 99.5% the speed of light, then by eq. (31), the time dilation factor is: $t = 10\, t_0$, so that a one-year passage of time experienced by someone on the rocket will correspond to a 10-year period for their identical twin on the ground. Mysterious, it definitely is.

This is related to the so-called 'twin paradox', which has a very long and colorful history[47].

---

[46] Interestingly, Einstein gave his own very elegant proof of the Pythagorean theorem, using the notion of similar triangles.

[47] See https://en.wikipedia.org/wiki/Twin_paradox

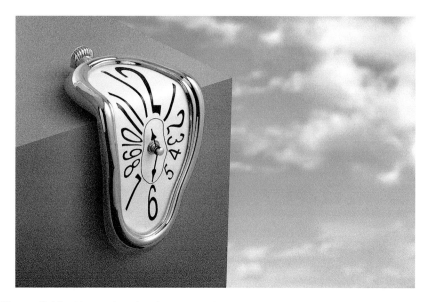

Figure 7.15: Since Einstein, time is no longer considered immutable, as was once thought.

Returning now to GPS, as we have seen earlier clocks at higher elevation run faster than those lower down. In fact, due solely to the altitude of the GPS satellites above the Earth, their atomic clocks would run about 45.8 millionths of a second per day faster than their counterparts on the ground.

On the other hand, the satellites' velocity, as we have just illustrated with time dilation, results in the orbiting clocks running about 7.2 millionths of a second per day slower than a clock on the ground. Overall, combining the two effects, the satellite clocks will run about 38.6 millionths of a second per day too fast compared to a stationary Earth clock. While this does not sound like much, if the effects of Relativity were not taken into account then errors of more than 10 kilometers (6 miles) *per day* would accumulate, rendering any such GPS system useless[48].

What needs to be done is now clear. The ticking rate of the satellite clocks needs to be slowed down by just this amount before they are even launched into space in order to match atomic clock time on Earth. Once they are in space, the

---

[48] Indeed, an error of 50 nanoseconds in the GPS timing system corresponds to an error of 15 meters in position (50 ns is the time it takes light to travel 15 m). So, 38 millionths of a second being 38,000 ns gives an error of:

$$\frac{38,000}{50} \times 15 = 11,400\,m.$$

The good news is that currently 1-cm GPS accuracy is now possible with suitable equipment.

two effects of Relativity – speeding up due to gravitational time dilation, and slowing down due to the velocity time dilation – will be compensated for and they will run in synchronization with Earth based atomic clocks.

So, the next time you use your GPS or cellphone to get you to a destination, spare a thought that the technology behind it is making use of Einstein's two very profound theories concerning the nature of time and space. German mathematician Hermann Minkowski summed up these ideas in a lecture he gave in September 1908: "Henceforth, space by itself and time by itself are doomed to fade away into mere shadows, and only a kind of union of the two will preserve an independent reality[49]." See Appendix II, where it is shown how Minkowski united space and time into Minkowski Spacetime.

Strikingly, a similar relationship as for time dilation as in eq. (31) applies to the *mass* of an object moving at a velocity v. If the initial stationary mass is m_0, then the formula for the moving mass, relative to its stationary one, is given by

$$m = \frac{m_0}{\sqrt{1 - v^2/c^2}}.$$

Again, when $v = 0$, then $m = m_0$, and otherwise one can see that as v increases, getting ever closer to the speed of light c, the mass of the object becomes increasingly large approaching infinity. This result has also been experimentally verified in particle accelerators when atomic particles are sped up to relativistic velocities. Of course, you are not going to notice this mass increase in your everyday experience.

Trying to rationalize the two great theories governing the very small and very large, namely, Quantum Mechanics and Relativity, is a crucial theoretical quest occupying the minds of many scientists. What is needed is a theory of quantum gravity, one that provides a mathematical theory of gravity, yet is in accordance with the rules of Quantum Mechanics. In the latter, particles are described by probabilities and waves, but not so for stars and planets. This turns out to be a tall order. "All approaches to the problem of quantum gravity agree that something must be said about the relationship between gravitation and quantized matter"[50].

String Theory, in which point-like particles are replaced by one-dimensional 'strings', is one such approach, as is M-theory involving membranes, of which strings become two-dimensional slices. In addition, holographic theories that have emerged from String Theory – as mentioned in Chapter 1 – that arise from the anti-de Sitter model are also playing an important role in coming to grips with a quantum theory of gravity.

---

[49] Address to the 80th Assembly of German Natural Scientists and Physicians, (Sep 21, 1908).

[50] 'Quantum Gravity', Stanford Encyclopedia of Philosophy, 2015.

RADIATION

Another interesting prediction of the Theory of Relativity is that the frequency of any radiation, say for example radio waves or light waves emanating from a dense star, will be altered by gravity. As mentioned previously, this was demonstrated with the frequency shift of radio signals from the Cassini spacecraft, as the radio waves passed near the Sun.

This happens because the waves must climb out of a 'gravity well', causing them to lose energy. Since the waves cannot slow down because the speed of light remains constant, this results in a decrease in the frequency of the waves ($E = hf$), shifting the waves towards the increased wavelength end of the spectrum[51]. This is known as a *gravitational redshift*, much like the redshift that was previously encountered due to the recessional velocity of a galaxy. The two redshifts are separate effects however, one due to gravity, one due to velocity (see Fig. 7.16).

So, let us look to the Heavens once again to find evidence of a gravitational redshift. In this instance, the redshift z is proportional to the mass of a star divided by its radius, i.e.,

$$z \propto \frac{M}{r}.$$

What is desired is a star of large mass M and small radius r in order to maximize the observable gravitational redshift z. Just such stars happen to be *white dwarfs*, the collapsed remnants of aging stars that are extremely dense, typically with a mass of our Sun but condensed into the size of the Earth.

In 1925, American astronomer Walter S. Adams attempted to observe this gravitational redshift of light by observing the white dwarf star Sirius B, the nearest white dwarf, which is a mere stone's throw away at 8.6 light-years[52]. This is the companion to the famous Dog Star, Sirius, the brightest star in the night sky.

But alas, observational and theoretical errors at the time meant that this line of confirmation had to wait until Daniel Popper in 1954, who determined the gravitational redshift of a different white dwarf star, 40 Eridani B. Popper found that the observed gravitational redshift "agrees with the predicted value within the uncertainties of both…"

The actual formula from Relativity relating gravitational redshift to the mass and radius is a bit of a handful, so instead we will use a simplified version:

$$cz = 0.635 \frac{M}{R}, \qquad\qquad (32)$$

---

[51] Recall, $E = hf$, so that f decreases as E decreases, and $\lambda = c/f$ implies that as frequency decreases, wavelength increases, becoming redder in the case of visible light.

[52] J.B. Holberg, Sirius B and the gravitational redshift, *J. Hist. Astron.*, Vol. 41, pp. 41−64, 2010.

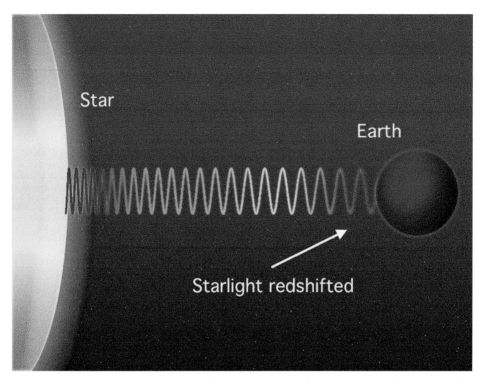

Figure 7.16: An exaggerated impression of light leaving a massive body, like a star, and the redshift of the light due to the star's gravity, as seen on Earth. (Image in public domain.)

where M is measured in terms of our Sun's mass (solar masses) and R is measured in terms of the Sun's radius[53].

Of course, we need to know the mass of the star and its radius as accurately as possible. In the case of 40 Eridani B, it is 16 light-years away and only slightly bigger than the Earth. Now there is some very good data for the mass and radius of our white dwarf (better than was available to Popper), and putting in these values of $M = 0.573$ solar masses, $R = 0.0136$ times the solar radius, we obtain from eq. (32)

$$cz = 26.75 \, km/sec.$$

---

[53] The redshift value is often given in terms of 'cz' in Astronomy and in other circumstances is a measure of recessional velocity, as mentioned above.

This is what *Relativity Theory predicts* given the measured values of M and R. An *actual* gravitational redshift obtained from examining the spectrum of 40 Eridani B has been measured at[54]:

$$cz = 26.5 \ km/sec,$$

giving very good agreement with the calculated gravitational redshift.

Thus, we have verified another aspect of the astounding Theory of Relativity, namely that mass not only bends the path of light but increases its wavelength, making objects look redder.

Since the gravitational redshift due to Relativity has been verified, this means that eq. (32) can also be used to calculate the mass M of a star, knowing its gravitational redshift value cz and radius R.

SYMMETRY AND GROUPS

> *Next to the concept of a function, which is the most important concept pervading the whole of mathematics, the concept of a group is of the greatest significance in the various branches of mathematics and its applications...*
> Russian mathematician Pavel Sergeyevich Alexandrov

One important aspect of many physical and biological theories is that of *symmetry*. Symmetry of form is very familiar to all of us, from the bi-lateral symmetry of the human form and a variety of other animals, to the shape of an orchid flower or the beauty of a snow crystal (Fig. 7.17).

Symmetries are also at the core of the Physics of our Universe. One such symmetry would be that of *time translation*; namely that the laws of Nature (such as Newton's Laws of Motion) are independent of time in the sense that the laws today were the same a million years ago and will be the same a million years hence. Even if we are not still around then.

There is also a symmetry of *spatial translation*, in which the same laws of Nature are as valid in New York as in Copenhagen, or in any city on a planet found in another galaxy. Nor do the laws of Nature change from day to night due to the rotation of the Earth. This is why Physics and life as we know it can exist in the first place.

---

[54] Value taken from: D. Koestler & V. Weidemann, On the mass of 40 Eri B, *Astron. J.*, Vol. 102, pp. 1152−1155, 1991.

Figure 7.17: Nature is all about symmetry, from the quantum level, to every day physical reality, to the Universe at large. (L) The bilateral mirror-image symmetry of an orchid flower (*Neomoorea wallisii*), and (R) a snowflake. The latter also has rotational symmetry, as a rotation by any integer multiple of $60°$ leaves the figure unchanged. The elaborate structure of snow crystals can be modelled via Cellular Automata (Chapter 5). (Images courtesy of (L) Ruud de Block; (R) Wilson Bentley collection.)

Symmetry in Mathematics plays a fundamental role in Physics, so let us try to come to grips with the most basic concept, a group. Consider a simple square, as in Fig. 7.18 far-left. If we rotate the square by $0°$, $90°$, $180°$, $270°$, we end up with the same square figure[55]. Each rotation was a physical operation performed on the original square, so let us give a name to each rotation operation, such as: r_0, r_1, r_2, r_3, respectively.

But there are four other *mirror reflection* operations that will leave the square unchanged (i.e., mirror-image symmetry). We could reflect the square about a horizontal axis line across the middle, or reflect across a vertical axis line through the middle as in Fig. 7.18 (bottom row left). Let us call these two further operations: r_4 and r_5.

Further, we still have two more operations that would leave the square invariant, namely reflecting the square through the diagonal from bottom left to top right or through the diagonal from bottom right to top left. Let us call these final two reflections: r_6, and r_7. The reflections: r_4, r_5, r_6, r_7, can also be visualized as flipping the square out of the plane about the respective axes. Note that each reflection operation, applied twice, returns the figure to its original position.

---

[55] Rotating the square by $360°$ is considered the same as a $0°$ rotation.

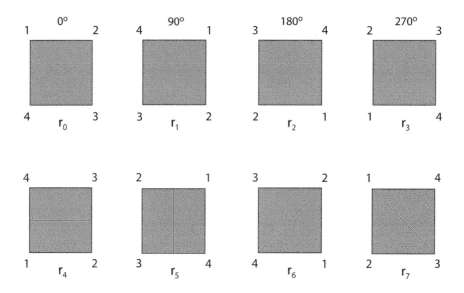

Figure 7.18: The four rotation operations (top line) and four reflection operations across an axis (bottom line) that can be performed on the initial square (top left) that leave the square invariant. (Illustration courtesy of Katy Metcalf.)

These eight operations, of four rotations and four reflections performed on the square, we can collect up together, and we call it a *group*. This specific group has a fancy name: the *dihedral group of order 8*.

$$G = \left\{ r_0, r_1, r_2, r_3, r_4, r_5, r_6, r_7 \right\}.$$

Actually, in Mathematics the term 'group' has a precise meaning. Indeed, we note that we can perform one operation right after another one. If we perform, say, r_1 and then r_2, then we will get r_3. We can write the sequential performance of operations just using juxtaposition: $r_2 \, r_1 = r_3$. Similarly, the reader can verify, with a square piece of paper labelled as in Fig. 7.18, that $r_4 \, r_7 = r_1$. Indeed, any composition of the eight operations will yield another member of our group.

This means that our group G is a *closed* system. If two giraffes mate, they do not produce an elephant, they produce another giraffe. Same with the elements of G. When they combine they form another element of G.

There is one basic technical property to define a group properly, and that is the *associative* property for combining elements; namely,

$$\left(r_i \, r_j \right) r_k = r_i \left(r_j \, r_k \right),$$

which is just the way addition and multiplication work with ordinary real (and complex) numbers. For example, $(2 + 3) + 4 = 2 + (3 + 4) = 9$, and likewise for the addition of any three real numbers.

Observe also that the operation r_0, which does no rotation at all, when composed with any other rotation

$$r_i r_0 = r_i = r_0 r_i,$$

leaves the initial rotation intact. This is just like addition with the number 0, such as: $7 + 0 = 7 = 0 + 7$. The number 0 is an identity in ordinary arithmetic, leaving numbers unchanged under addition, and so r_0 is also called an *identity* of our group as it serves the same function. Similarly, the number 1 is an identity when multiplying by any real (or complex) number, as $7 \times 1 = 7 = 1 \times 7$.

Finally, note that for, say the rotation r_3, if we compose it with r_1, the result will be the identity, r_0. Indeed, for any operation r of our group G, there is always going to be some other member in G denoted by r^{-1} that, when composed with r, will yield the identity. That is,

$$rr^{-1} = r_0 = r^{-1}r.$$

This other member is called the *inverse* of the first one, and the first one is also the inverse of the second one; that is, they are inverses of each other[56].

This is certainly also the case in arithmetic with the operation of addition, as $7 + (-7) = 0 = (-7) + 7$. Likewise, any real number n has an additive inverse $(-n)$ [57].

So now we have all the criteria that make up a properly defined group. It should have some binary operation to combine its elements; the group should be a closed system when combining its elements; the elements of the group should obey the associative property; the group should have an identity member that leaves all others unchanged; and finally, each element should have an inverse in the group that yields the identity when the two are combined under the group's operation.

The preceding operations on a square form one main type of group, as it has a finite number of elements, namely 8. In a similar fashion, we could form another group of operations that act upon the snowflake in Fig. 7.17, namely six rotations

---

[56] The inverse can be referred to as *the* inverse as it is unique for each element of G. The proof follows from the associative property of the group.

[57] Indeed, the set of real numbers, $\mathbb{R}$, form a group under the operation of addition. But $\mathbb{R}$ together with the operation of multiplication has no multiplicative inverse for the number 0 and so the reals do *not* form a group under multiplication. However, for every real number $n \neq 0$, there *is* a multiplicative inverse, n^{-1} such that $n \cdot n^{-1} = 1 = n^{-1} \cdot n$. Similarly for the set of complex numbers, $\mathbb{C}$, every $z \neq 0$ has a multiplicative inverse, z^{-1}, according to how we defined multiplication of complex numbers in Chapter 3. Sets such as $\mathbb{R}$ and $\mathbb{C}$ form what are called *fields*.

and six reflections that leave the snowflake invariant[58]. This set of operations would form another group consisting of 12 elements[59].

But other groups of operations can have infinitely many elements. Instead of a square, let us simply take a circle, specifically the circle of radius = 1 in the complex plane. From our discussion of complex numbers, we know that every point on the unit circle can be represented as: $z = e^{i\theta}$, since the modulus of z is $r = 1$ (Fig. 3.4, Chapter 3).

Moreover, each point $z = e^{i\theta}$ on the unit circle actually represents a rotation along the circle of the point $z = 1$ by the angle θ, since θ is the argument of z. We can combine two such rotations, $e^{i\alpha}$ and $e^{i\beta}$ via the operation of multiplication:

$$e^{i\alpha} \cdot e^{i\beta} = e^{i(\alpha+\beta)},$$

merely using the rules for multiplying numbers with exponents[60]. The result is another point (rotation) on the unit circle whose argument is the sum of the two arguments, namely $\alpha + \beta$ (Fig. 7.19).

Therefore, combining two points on the circle via multiplication reduces to nothing more than the addition of their arguments.

Now, by taking the set of all points (rotations) of the unit circle, which we denote by $\mathbb{T}$ (because the circle is effectively a skinny Torus, i.e. donut), we again have a well-defined group under the operation of multiplication. The elements of $\mathbb{T}$ are *closed* under the operation of multiplication as seen above, and the *associative* property is valid because of the associative property for the numbers involved in the exponents of each point $e^{i\theta}$. Furthermore, any point (rotation) on the unit circle can be combined with an *identity* rotation given by $e^{i0} = 1$, (which amounts to a rotation by zero radians, which is no rotation at all) so that:

$$e^{i0} \cdot e^{i\theta} = e^{i\theta} = e^{i\theta} \cdot e^{i0}.$$

Finally, each rotation, $e^{i\theta}$, has an *inverse*, $e^{-i\theta}$, whereby:

$$e^{i\theta} \cdot e^{-i\theta} = e^{i0} = e^{-i\theta} \cdot e^{i\theta},$$

and $e^{i0} = 1$ is the identity under our multiplication operation.

---

[58] The six reflections consist of three reflections about axes connecting opposite vertices and three reflections about axes connecting the midpoints of opposite sides of the hexagon formed by the snowflake.

[59] In general, any regular polygon having n sides will have n rotational symmetries and n reflection symmetries, and the set of symmetries will form a (dihedral) group having $2n$ elements.

[60] $e^x e^y = e^{x+y}$, for both real numbers and complex numbers and clearly the multiplication is also commutative.

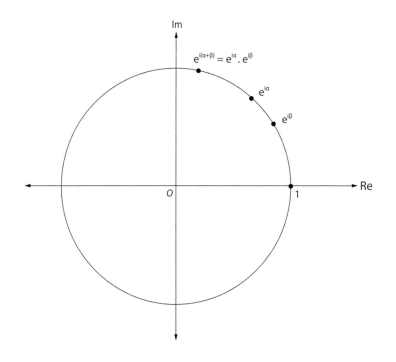

Figure 7.19: Combining two rotations of the circle group $\mathbb{T}$, namely $e^{i\alpha}$ and $e^{i\beta}$, via the operation of multiplication produces the rotation $e^{i(\alpha + \beta)}$. (Illustration courtesy of Katy Metcalf.)

The only technical thing we need to bother with is what happens when we run over 360° when combining rotations. The answer is that we simply subtract 360°, (actually, 2π, since we are dealing with radians) and we are back in the group game. Just like when it is 10 o'clock and three hours later it is 1 o'clock, not 13 o'clock (on a standard 12-hour clock face, of course), so we are all very used to this adaptation.

The *circle group* $\mathbb{T}$ is called a *continuous group*[61] as opposed to the finite *discrete group* discussed for the square. An infinite group that also happens to be discrete is the set of integers with the operation of addition. The number 0 is the identity, and every number n has an inverse, namely $-n$. We all learned in school that addition is associative, so that the integers under addition form a discrete group denoted by $\mathbb{Z}$.

---

[61] The members of the group $\mathbb{T}$ change *continuously* with the angle θ, but we need not define the term 'continuous' mathematically rigorously. A dictionary interpretation suffices.

Another continuous group would be the collection of all spatial translations of a region in space. For example, taking any region R of the plane and displacing every point in the region by the magnitude and direction of a fixed vector v, one has a translation of the region which we denote by T_v. Translating the original region R once again, say by a different vector u this time, gives the translation T_u.

Then there is nothing to stop us from performing the first translation (T_v) of R followed by the second translation (T_u) of the T_v-translated region, which we can denote by the composition: $T_u T_v$. This sequential performance of the translation operations will define a group **T** of all such vector translations of R, with the inverse given by moving the region backwards by T_{-v}, and the identity is a translation by the zero vector. Like the circle group, the group **T** will be continuous, as both the length and direction of each vector can be varied continuously.

Although a mathematical group seems to be an abstract construct, it turns out that much of Quantum Mechanics has evolved into the theory of groups, with many books being written on the subject[62].

Initially this was not a welcome development, with the not uncommon term *Gruppenpest* expressing the prevailing sentiment of many early physicists. The plaintive expression "*Gruppenpest* must be abolished" is attributed to Schrödinger.

Groups also play a role in Chemistry and Crystallography, and even in data encryption, such as in the famous public-key RSA encryption[63]. Indeed, groups are a major component of the mathematical sciences.

Let us go back to the notion of symmetry that we started our discussion of groups with. Symmetry plays a whole host of different roles in the physics of the Universe. At the core of this deep connection is the work of German mathematician Emmy Noether (1882−1935)[64]. She has been hailed as the finest female mathematician ever. Noether's father was also an outstanding mathematician, but his daughter's talent far eclipsed his own.

After receiving her PhD degree from Erlangen University in 1907, Emmy Noether worked there without pay for many years, because prejudice against women working at universities was rife. It was the great David Hilbert and his colleague Felix Klein at the University of Göttingen who attempted to have her join them on the staff.

---

[62] Beginning with Eugene Wigner's, *Gruppentheorie und ihre Anwendung auf die Quartenmechanik der Atomspektiren,* Braunschweig: Friedr. Vieweg, 1931. The 1959 English translation and expanded version: *Group Theory and its Application to the Quantum Mechanics of Atomic Spectra*, is still available.

[63] RSA encryption is due to Ron **R**ivest, Adi **S**hamir, and Leonard **A**dleman of MIT in 1978. It was essentially discovered by Clifford Cocks of British Intelligence some years earlier but remained classified.

[64] Pronounced NUR-tuh.

They were working on Einstein's newly published General Theory of Relativity at the time (1915), and Noether had valuable mathematical expertise which they sought. Göttingen during this period was the pre-eminent center for Mathematics in the world[65], but the prospect of her appointment met with sexist vocal opposition from the faculty senate: "What will our soldiers think when they return to the University and find that they are expected to learn at the feet of a woman?"[66] This pales in comparison to other comments made about her dress and appearance, as well as other derogatory remarks describing her as *Der Noether*.

Nevertheless, Emmy Noether did come to Göttingen and taught classes that were officially assigned to Hilbert. Subsequently, in 1922, the university granted her a purely honorary title, and later she received a small stipend. Noether, who was Jewish, was forced by the Nazis to leave her position at Göttingen and she took up a position at Bryn Mawr College in Pennsylvania in 1933. Less than two years later, she died from complications after surgery, aged 53.

Paying homage to another great mind, her obituary in the *New York Times* was written by no less than Einstein himself.

Confronting the problem that Hilbert and Klein were grappling with concerning Relativity (that of conservation of energy), Noether not only resolved it, but changed the course of Physics with a now famous theorem that bears her name.

To understand the result now known as *Noether's theorem*, we first must address the notion of conservation laws in Physics. There are several conservation laws, which basically mean that certain physical quantities remain unchanged over time. For example, conservation of energy means that energy can manifest in different forms (as in $E = mc^2$), but can never be destroyed as its total value always remains invariant.

There is also conservation of (linear) momentum (mass times velocity: mv) of an object moving in a straight line, meaning its momentum remains unaltered if free of all external forces, and similarly for the conservation of angular momentum of a rotating object. In addition, there is conservation of electric charge whereby, in an isolated system, the total electric charge remains constant over time even if there are interactions within the system. These conservation laws are the foundations upon which the Physics of our Universe is built. But they do not seem to have anything to do with symmetry. Enter Emmy Noether.

---

[65] All this changed with the rise of Nazism in the 1930s. When Hilbert was asked about the state of Mathematics at Göttingen "now that it had been freed of the Jewish influence," by the Minister of Education who was newly appointed by the Nazis, he replied, "Mathematics in Göttingen? There is really none anymore." (From: *Hilbert*, p. 205. See Bibliography).

[66] *Ibid*, p.143.

Noether's theorem is based on continuous symmetry groups and expresses the fact that:

Each of the conservation laws of Physics is associated with an underlying symmetry property in Nature.

That is, where one finds a conservation law there will be a corresponding symmetry property in Nature, and vice versa. This gives physicists a guiding principle when they find one or the other. The symmetries involved are 'continuous', such as in the rotations of the unit circle, or as in the symmetry of spatial translations, discussed above.

At the heart of the correspondence between conservation laws and symmetry is a fundamental inclination of the Universe: the so-called, 'principle of least action'. This is conceptually equivalent to the reader avoiding cutting the grass or washing the dishes. Nature does not like to expend unnecessary energy any more than you do. Photons in a beam of light will travel in a straight line, as this is the shortest path between two points and thus minimizes the time travelled.

Noether's theorem pairs up conservation laws with symmetries by invoking the principle of least action in a very mathematically precise manner, but the details are beyond the scope of this book.

Taking the conservation of energy, for example, energy is conserved as a consequence of the symmetry of time translation. Linear (angular) momentum is conserved because of the symmetry of spatial translation (rotation). These conservation laws are fundamental to establishing the validity of the equivalence of energy and mass via Einstein's equation $E = mc^2$ discussed earlier in this chapter.

At the quantum level, *Supersymmetry* (quaintly known as SUSY) is a theory in Physics that proposes a dual particle (a superpartner) for each particle that is found in either of the two main classes of particles, *bosons* (force carrying) and *fermions* (mass carrying). As we have seen, subatomic particles have a property called *spin,* with those in the boson class having integer spin values. For example, a photon has spin 1 and the Higgs boson has spin 0[67].

The other class of particles, the fermions, have half-integer spin, like 1/2, 3/2, …, so for example, the electron has spin 1/2. The SUSY hypothesis supposes that each particle in one class has an equivalent particle in the other class. So far there has been no experimental evidence supporting the hypothesis, but there are deep theoretical reasons why it might be true[68]. Within a SUSY model there may exist a conservation law in accordance with Noether's theorem, and it has been suggested that in such a case, SUSY could provide a supersymmetry particle candidate for the unseen dark matter in the Universe.

---

[67] The discovery of the Higgs particle arose out of theoretical considerations of the notion of *symmetry breaking*.

[68] In this context it should be mentioned that supersymmetry is one of the requirements of String Theory.

8

The Unknowable Universe

There are more things in heaven and earth, Horatio, than are dreamt
of in our philosophy...
Shakespeare's *Hamlet*[1]

Some things are intrinsically unknowable, and no amount of mathematical analysis will reveal their nature.

GÖDEL INCOMPLETENESS

Most people think that any mathematical problem can be solved eventually, even the most difficult, like the Riemann Hypothesis. The difficult problems may take 100 years or more, but eventually it is thought they will be solved by some very brilliant mathematician because they are inherently solvable. Unfortunately, this is not the case, and this was demonstrated by the Austrian mathematician Kurt Gödel in two theorems published in 1931 that delineated the limits of formal mathematical systems.

What Gödel showed was that there are certain statements within a formal system that can neither be proved nor disproved using the axioms and rules of that system. That is, such statements are 'undecidable' within the system. Gödel not only showed that such statements (propositions) can exist, but actually produced one that was undecidable within the formal system in question. Indeed, Gödel

---

[1] We use the word 'our' as it appears in the First Folio of 1623, but 'your' appears in the second quarto of 1605, so either is appropriate but with slightly different meanings.

© Springer Nature Switzerland AG 2020
J. L. Schiff, *The Mathematical Universe*, Springer Praxis Books,
https://doi.org/10.1007/978-3-030-50649-0_8

produced a statement *G* that was a formalized expression of its own unprovability. Gödel then proved *G* to be true, demonstrating that it was unprovable.

We have already encountered one such example, the Continuum Hypothesis (CH), which cannot be proved nor disproved within the confines of axiomatic set theory (ZFC).

This is not such a strange phenomenon, as it is analogous to the interesting self-referential statement:

This sentence is false.[2]

If we assume the sentence to be true, then it must be false. But if it is false, then it must be true. This kind of paradox is what inspired Gödel to create a similar situation in a mathematical setting.

HALTING PROBLEM

There have been other ramifications of the Gödel incompleteness theorem. One such was found by computer pioneer Alan Turing in 1936, when he showed that there is no procedure to determine whether a computer program will eventually complete its task (halt) or continue to run indefinitely. Turing's proof was one by *reductio ad absurdum,* in that one assumes that there is such a procedure and derives a contradiction along the lines of the Cantor diagonal method[3].

This is known as the *halting problem*, and Turing demonstrated that there is no such general procedure for deciding whether the computer program ever halts or not. Thus, the question is undecidable.

That is a pity, for if the halting problem could be solved, then there are a lot of conjectures concerning numbers for which it could be decided whether or not they are either true or false. For example, there is the *Goldbach Conjecture*, made by Christian Goldbach (1690–1764) in a letter to Leonhard Euler in 1742, which says that every even number greater than 2 can be derived from the sum of two primes. Thus, for example,

$$16 = 11 + 5, \qquad 58 = 53 + 5, \qquad 673248 = 103787 + 569461.$$

Euler was sure the proposition was true but was not able to prove it. Nor has anyone else been able to for that matter, up to the present day. Clearly, one can write a simple computer program to check the conjecture, up to a point. Just start with 4, 6, 8, … , and check the sums of the smaller primes to see if they equal the number in question. In fact, this is not such a bad idea if you think the conjecture might

---

[2] This is one version of the so-called 'liar paradox' which dates back to the ancient Greeks. It has been the subject of much discussion ever since.

[3] See Appendix III, where the Cantor diagonal method is employed. It should be noted that a proof of the halting problem does not necessarily require the method of contradiction.

be false. Alas, it seems that even numbers up to 4×10^{17} have been verified by a computer and every single one was equal to the sum of two primes[4].

What Turing showed was that if we wished to let our computer continue to run, then there is no way to decide *a priori* whether it will ever halt or not. So, a computer is not very likely to settle the conjecture, since if Euler thought the conjecture was true, then it probably is true, and therefore our computer program would never halt[5].

EMX

On the other hand, computers are getting smarter and smarter and learning to do all sorts of tasks, from playing chess to making medical diagnoses. Of course, many of the techniques used to achieve 'machine learning' involve a lot of Mathematics, as the computer goes through a period of training. This often involves minimizing the discrepancy between what has already been learned by the machine and the desired outcome. We saw an example of this earlier, in which a fitness test was introduced to train a cellular automaton to learn how to arrange a random selection of black and white cells into a checkerboard pattern (Fig. 5.8 in Chapter 5). Similarly, a diagnosis of food poisoning might be close, but not quite close enough if the problem is actually appendicitis. Therefore, like a real doctor, a computer doctor must hone its skills and learn from experience.

Here is another real-life problem for a computer. Suppose you have a website that allows advertisements to be posted on it, and you wish to maximize the number of hits the ads receive. Each of the large pool of available ads targets a specific set of individuals, such as movie buffs, people seeking to lose weight, dog lovers, etc., The goal is to post ads that result in receiving the most visits, but without knowing in advance what sort of people will visit the website. The problem, therefore, is to choose the correct ads in order to maximize the number of viewer-hits and to get a computer to learn how to do this. Sounds straightforward enough.

This problem in machine learning was tackled by a group of scientists who labelled it 'EMX', for 'estimating the maximum'[6]. What the investigators found was rather remarkable:

"We describe simple scenarios where learnability cannot be proved nor refuted using the standard axioms of Mathematics. Our proof is based on the

---

[4]This was done by Tomás Oliveira e Silva of University Aveiro, Portugal, and the result was double-checked in 2013.

[5]There is also Goldbach's 'weak' conjecture, which states that every odd number greater than 5 can be written as the sum of three primes. A proof of this conjecture was given by mathematician Harald Helfgott in 2013.

[6]S. Ben-David, *et al.*, Learnability can be undecidable, *Nature Machine Intelligence*, Vol. 1, pp. 44–48, 2019.

fact the continuum hypothesis cannot be proved nor refuted. We show that, in some cases, a solution to the 'estimating the maximum' problem is equivalent to the continuum hypothesis."

Since CH cannot be proved or disproved within the standard axioms of set theory, then whether or not the EMX problem can be solved is itself undecidable. Something so absolutely abstract as whether or not there is an infinite cardinal between $\aleph_0$ and the continuum $\mathfrak{c}$, meets head on with a computer problem in real life. Just brilliant.

WHERE IS IT, DR. HEISENBERG?

Part of the quantum mechanical description of the world is a famous little formulation derived by Dr. Werner Heisenberg (1901–1976) that quantifies what is actually knowable about the world at its smallest levels. Namely,

$$\Delta x \Delta p \geq h/4\pi,$$

where x represents the position of the particle in space, p is its momentum, and h is the Planck constant[7].

Again, we find π creeping into the most arcane considerations. Indeed, it is ubiquitous in Physics. The delta symbol, Δ, represents the intrinsic *uncertainty* in the position, and momentum, respectively. What the formula is saying is that for any measurement involving a particle's position and momentum, as the uncertainty in one component goes down, the other must go up correspondingly (Fig. 8.1).

It must be appreciated that the Uncertainty Principle is not a statement about our current technological ability to measure such properties. Rather, it is an inherent property of Quantum Mechanics due to the wave nature of reality. However, if you cannot find your keys, you cannot blame it on Heisenberg, as absentmindedness is not part of the principle.

For particles with mass, such as an electron, the uncertainty involved would be between position and velocity, since its momentum is: $p = mv$, and the mass is already known. As we have seen in Fig. 7.3 in Chapter 7, the position of a particle is uncertain and only a probability of finding it in a particular position can be given.

One consequence of the Uncertainty Principle lies in the nothingness of space itself. The vacuum of empty space comprising most of the Universe consists of fields that have a *vacuum energy* or *zero-point energy*, although just how much energy a volume of space contains according to Quantum Mechanics is still inconclusive. Better results are obtained by invoking the Theory of Relativity, which gives a value very close to zero, and this vacuum energy pervading all of space could be accountable for the dark energy thought to be responsible for the apparent accelerating expansion of the Universe.

---

[7] Sometimes Planck's constant h is replaced by $\hbar = h/2\pi$.

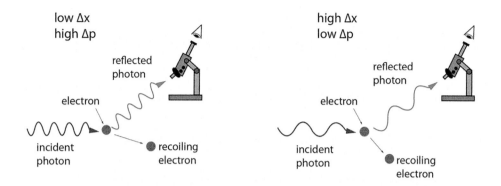

low Δx
high Δp

high Δx
low Δp

reflected
photon

reflected
photon

electron

electron

incident
photon

recoiling
electron

incident
photon

recoiling
electron

Figure 8.1: An illustration of how attempting to observe a particle such as an electron disturbs its position and velocity. (L) Using higher frequency photons (shorter wavelength) yields a more accurate determination of the electron's position. However, as more energy is imparted to the electron ($E = hf$), greater uncertainty is produced in the momentum of the recoiling electron. (R) Lower frequency photons yield greater uncertainty in the electron's position but a more accurate determination of its momentum. (Illustration courtesy of Katy Metcalf.)

Recall that the amount of matter in the Universe is virtually at the critical density $\rho_c \approx 10^{-23}$ gm/m^3, and that about 69% of this was found by the Planck Space Telescope to be in the form of dark energy. Since $E = mc^2$, this dark energy converts to a mass equivalent of $\sim 7 \times 10^{-24}$ gm/m^3, which generates gravity and hence a (negative) pressure contributing to the expansion of the Universe.

This dark energy contribution to the expansion of the Universe is denoted by Λ, in Einstein's field equations (eq. 30). Its value is so small as to be practically zero, but if it were larger, the Universe would expand too rapidly and stars and galaxies would not have been able to form, and neither would we.

While the nature of dark energy is very much a mystery, the fate of our Universe depends upon this very entity we know virtually nothing about.

In fact, as the Universe expands and the volume of space increases, the totality of dark energy also increases as it is an intrinsic property of space itself. However, matter and radiation are not increasing overall and are thus becoming more and more diluted as the Universe expands. What this means is that the expansion of Universe will become increasingly dominated by dark energy and, with little to hinder its effects, the expansion should accelerate. Strong evidence for this acceleration has already been found[8].

Therefore, the ultimate fate of our Universe could conceivably be dark indeed, and it is seemingly all down to that one little constant Λ in eq. (30).

---

[8] Scientists Saul Perlmutter, Brian P. Schmidt, and Adam G. Riess won the 2011 Nobel Prize for Physics, *"for the discovery of the accelerating expansion of the Universe through observations of distant supernovae."*

Indeed, the symbols of Mathematics are like a universal alphabet that, when put together just so, tells us something about how the world works. From the Pythagorean Theorem, $c^2 = a^2 + b^2$, to $E = mc^2$ two and a half millennia later, what a great blessing it is that Mathematics has been bestowed upon us and what a great mystery that this should be so. These little squiggles on paper are like notes of music, and if you can read the score you can hear the Music of the Spheres.

SUMMING UP

What we have considered in this monograph is but the tiniest inkling of how Mathematics is used every day by scientists to uncover the workings of our world, from the sublime vastness of the Universe itself down to the ridiculously small quantum behavior of subatomic particles.

This understanding of the Universe, at both ends of the distance scale, rests upon the twin pillars of 20th century Science: the Theory of Relativity and Quantum Mechanics (QM). However, there are a number of incompatibilities between General Relativity and QM. As already mentioned in the preceding chapter, Paul Dirac had already united Special Relativity with Quantum Mechanics, so the real problem lies with QM and gravity, as formulated in the General Theory of Relativity.

At the most fundamental level, spacetime is a geometric continuum in General Relativity, whereas QM deals with discrete quanta as well as uncertainty and probabilistic outcomes. Therefore, the areas of incompatibility are many and the problem of reconciling the two magisterial theories is a formidable one for the 21st century.

These attempts fall under the heading of Quantum Gravity and there is a myriad of such theoretical avenues being pursued, with String Theory being just one of many. It should be mentioned that one of the key ingredients of the String Theory approach is that it invokes the ideas of anti-de Sitter space discussed in Chapter 1.

A very readable account of Relativity, QM, and a rival to String Theory called Loop Quantum Gravity (LQG), can be found in the delightful book by Carlo Rovelli in the Bibliography. In LQG, spacetime is not a continuum that is infinitely divisible, but is granular, with grain areas on the order of the Planck length squared; that is $\lambda_p^2 \left(\sim 10^{-66} \ cm^2 \right)$. It is simply not possible to speak of an area that is smaller, nor a volume smaller than $\lambda_p^3 \left(\sim 10^{-99} \ cm^3 \right)$.

As such, LQG rules out a Big Bang scenario in which the Universe arose from a singularity of zero volume and infinite density. In LQG, there was instead a Big Bounce, in which our Universe arose from a former version of itself that had contracted down to a very small size, though not a singularity. On the other hand, as mentioned above, the current evidence suggests that the Universe is expanding at an ever-increasing rate and it will expand in this manner forever, so there is no going back. However, since dark energy is so little understood, perhaps its nature will change at some time in the distant future and the Universe will return to its origins.

Whatever form the reconciliation happens to take, assuming there even is one, it will almost certainly be written in the language of Mathematics. Hopefully, with a bit of luck, we will have billions of years yet to work out the details.

Let us conclude with the famous allegory narrated by Socrates in Book VII of Plato's *Republic*. Socrates imagines a cave in which prisoners have been chained up since childhood, facing the back wall of the cave so that they cannot even turn to see behind them. In the distance behind them is a burning fire, and in between the fire and the prisoners is a low wall, "like the screen over which marionette players show their puppets. Behind the wall appear moving figures, who hold in their hands various works of art, and among them images of men and animals, wood and stone, and some of the passers-by are talking and others silent."[9]

For the prisoners, the shadows being played out on the wall thus become their entire experience of reality (Fig. 8.2). The prisoners even make up names for the shadowy figures they see on the wall and voices from the passers-by will echo off the wall as if they come from the figures.

Interestingly, note the similarities between the allegory of Plato's cave and the Holographic Principle discussed in Chapter 1.

Figure 8.2:   An illustration of the scene in the cave from the allegory in Plato's *Republic*. For the prisoners, the shadows are what they infer to be a reality beyond. (Illustration courtesy of Katy Metcalf.)

[9] Text taken from the translation by Benjamin Jowett.

While the shadows being played out on the wall become the entire experience of reality for the prisoners, outside the mouth of the cave there is an entirely different reality, and what they witness on the wall is only a poor second-hand version.

Is Mathematics merely the fire that allows mankind to see the shadows of a more remote ultimate reality?

Appendix I

The important thing to remember about mathematics is not to be frightened …
Biologist Richard Dawkins

Everything should be made as simple as possible, but not simpler …
Albert Einstein

The following material is supplementary to the text, for those who will 'not be frightened' and wish to pursue some of the Mathematics alluded to.

BEING REASONABLE

Here is another example, based on the law of the excluded middle, which illustrates Brouwer's point about proofs by contradiction. Suppose our proposition *P* is that: *the number of prime numbers, 2, 3, 5, 7, 11, … is infinite.* Note that this is not intuitively obvious, because the prime numbers overall do become more sparse the higher we count, and so it is possible they could simply peter out. Let us see if they do. The details simply involve a bit of rigorous arithmetic.

Let us again assume that *P* is false; that is, *there are only finitely many primes.* If this were the case, we can make a list of all of them, so let us write down all the prime numbers, and let us call them: $p_1, p_2, p_3, \ldots, p_n$. This is our list of *all* the primes. It might be a very long list and it would take us a long time to write all of them down, but in principle, as the list is finite, we can make such a list.

What should we do with our list of all the primes now? Here is what Euclid thought of doing some 2,300 years ago. Why not multiply all of the primes together then take this product of prime numbers and just add the number 1 to it?

© Springer Nature Switzerland AG 2020
J. L. Schiff, *The Mathematical Universe*, Springer Praxis Books,
https://doi.org/10.1007/978-3-030-50649-0

This turns out to be a clever thing to do. The result will be a very large new number, so let us give it a name and call it q. Therefore,

$$q = \left(p_1 \cdot p_2 \cdot p_3 \cdots p_n \right) + 1.$$

Now we ask ourselves, what kind of number is q? There are two basic types of positive integers; namely the primes, and all the other non-prime numbers that are called *composite* numbers. The composite numbers can be divided (without remainder) by numbers other than 1 and themselves; they are numbers like 8, 10, 21, 100, and so forth. Thus, our new number q that we formed is either a prime or a composite number. The interesting thing about composite numbers is that they can be broken down (factored) into a product of primes.[1] So for example,

$$8 = 2 \cdot 2 \cdot 2, \qquad 10 = 2 \cdot 5, \qquad 21 = 3 \cdot 7, \qquad 100 = 2 \cdot 2 \cdot 5 \cdot 5.$$

The primes form the building blocks, or atoms, of the number system and any composite number will have some prime that divides it. This is important to note.

Now, our special number q that we formed above is either a prime or composite. There is no other option. In either case, *some* prime number must divide our number q.

As we have our list of *all* the prime numbers in hand, it must be that the prime number that divides q is one of the prime numbers in our list, which we will call $p = p_j$. We do not need to know explicitly what this number actually is, it is just some prime on the list.

However, by a simple fact of arithmetic, if p divides some integer n, and p also divides another integer m, then p will divide their difference $n - m$. So, for example, 3 divides 21 and 3 divides 9, so 3 also divides $21 - 9 = 12$.

Therefore, as the prime number p divides our number q, and certainly p divides the product that we formed earlier $(p_1 \cdot p_2 \cdot p_3 \cdots p_n)$ since it is one of the factors in the product, we can conclude that p must also divide the difference, which happens to have the value 1:

$$q - \left(p_1 \cdot p_2 \cdot p_3 \cdots p_n \right) = 1.$$

But p being a prime number *cannot* divide 1; that is nonsense. So here we have arrived at a contradiction by assuming that our proposition P is false. The classical conclusion is that our proposition P is true, and thus there are infinitely many prime numbers.

---

[1] This fact is known as the *fundamental theorem of arithmetic*, namely that every positive integer that is itself not prime can be decomposed into a product of prime numbers. The product happens to be unique, except for the order of the multiplication of the prime numbers.

What the preceding argument does not do is *construct* all these infinitely many primes, as they are just assumed to exist, somewhere out there in the Mathematical Universe. This is why Brouwer's adherents, known as *constructivists (or intuitionists)*, do not accept this type of proof. For them, the falsehood of non-existence (i.e. there are not infinitely many primes) does not provide an algorithm for existence, which is absolutely correct. This has led to a desire to produce mathematical proofs that are constructive in nature wherever possible.

Indeed, Euclid's original argument was actually *not* a proof by contradiction and with a slight variation of our approach, Euclid's direct proof ensues.

For comparison of the two methods, suppose that we have compiled that list of (finitely many) prime numbers. We denote them as before: $p_1, p_2, p_3, \ldots, p_n$, and again we form the number q,

$$q = \left(p_1 \cdot p_2 \cdot p_3 \cdots p_n\right) + 1.$$

If q is a prime number, then we have found a *new* prime that is clearly not in the list. On the other hand, if q is a composite number, then q as we argued above, must be divisible by some prime number, p.

Now we argue that this prime p *cannot be in our list*, for if it were, then p would obviously divide product, $(p_1 \cdot p_2 \cdot p_3 \cdots p_n)$. Since p is a prime that also divides q, then as in the argument above, p would divide the difference between q and the product, namely the number 1. Since no prime number can divide 1, we conclude that p cannot be included in our finite list of primes.

Thus, we have again found a new prime number (either q or p) not included on our list of primes. We conclude that any such finite list of primes must necessarily be *incomplete*, as we can always find one more prime that is not on the list. Moreover, we have (or rather, Euclid has) even provided an algorithm for finding this new prime number. Again, the number of primes must be infinite.

Appendix II

HYPERBOLIC GEOMETRY AND MINKOWSKI SPACETIME

We will consider the hyperboloid of two sheets (Fig. A.1) but only the top surface. All points (x, y, z) on this surface (blue) satisfy the equation[1]

$$x^2 + y^2 - z^2 = -1,$$

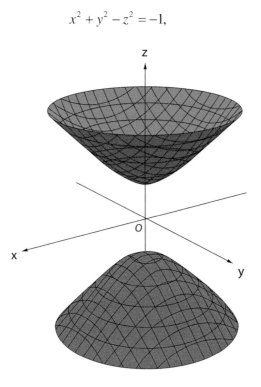

Figure A.1 The hyperboloid of two sheets, but we will only consider the top surface. (Illustration courtesy of Katy Metcalf.)

---

[1] The two-sheet hyperboloid can also be defined by: $-x^2 - y^2 + z^2 = 1$. The same two surfaces arise.

© Springer Nature Switzerland AG 2020
J. L. Schiff, *The Mathematical Universe*, Springer Praxis Books,
https://doi.org/10.1007/978-3-030-50649-0

for $z \geq 1$.

We are going to put a disk (in blue) of radius = 1 in the xy-plane with the hyperboloid surface directly above it, as in Fig. A.2. All the points on the surface of the hyperboloid can be projected down into this disk by taking a straight line from any point on the hyperbola and connecting it to the point $(0,0,-1)$. Where this line intersects the xy-plane is the unique projected point of the hyperboloid into the disk. In this sense, we have a representation of the entire hyperboloid surface completely contained inside the disk of radius = 1.

For any two given points on the hyperboloid, a 'hyperbolic straight line' (a *geodesic* – in red) connecting the two points on the hyperboloid, is represented by the points of intersection from a plane passing through those two points and the origin at $(0,0,0)$.

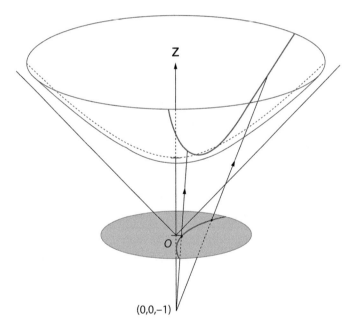

Figure A.2 A plane passing through two points on the hyperboloid and the origin determines a hyperbolic straight line (*geodesic* – in red) on the hyperboloid. Its counterpart in the Poincaré disk is the circular red arc obtained by projection as discussed in the text and depicted in Fig. 1.8 in Chapter 1. (Illustration courtesy of Katy Metcalf.)

All points on this hyperbolic straight line can be projected downward onto the disk below by drawing a Euclidian straight line from the point on the hyperboloid to the point $(0,0,-1)$ in the manner above. This determines a corresponding geodesic in the Poincaré disk – also in red. In this manner, we can transfer the geometry of the hyperboloid to an equivalent representation in the disk where it is easily visualized, as we saw in Fig. 1.8 in Chapter 1. Note that as a point on the hyperboloid moves further and further away from the point (0, 0, 1), the corresponding point in the blue disk is moving out to the boundary representing infinity.

By adjoining a time coordinate to three spatial coordinates, $x = (x, y, z, t)$ we have a point (or a vector) in a four-dimensional Euclidean space. We couple with that a way to define distance, which is somewhat analogous to that of Euclidean distance. For a point (x, y, z, t), we define its distance s (squared) from the origin by the equation

$$s^2 = x^2 + y^2 + z^2 - t^2 \tag{33}$$

Note that this is very similar to what would be the Euclidean distance (squared) to the point which is given by $s^2 = (x^2 + y^2 + z^2 + t^2)$, but for the negative sign, so the geometry is called *pseudo-Euclidean*.

Let us consider all vectors: $\mathbf{v} = (x, y, z, t)$ that satisfy the equation

$$x^2 + y^2 + z^2 - t^2 = -1. \tag{34}$$

Coupled with a means to measure distance, as in eq. (33), we obtain a model of *Minkowski Spacetime* that becomes a four-dimensional framework for Einstein's Theory of Special Relativity.[2]

This framework is a 'hyperbolic space' because by dropping one of the space coordinates, say z, then the surface defined by eq. (34), consists of all the points satisfying

$$x^2 + y^2 - t^2 = -1,$$

which is just a hyperboloid of two sheets as defined above.

Interestingly, if we substitute the hyperboloid given by eq. (34) into the distance expression of eq. (33), then we obtain the distance relation

$$s^2 = -1,$$

that is, $s = \sqrt{-1}$. This means that our space may be considered as a 'sphere' of radius $= i$. Now that is certainly interesting geometry.

---

[2] Normally, equations such as (33) involve the velocity of light c in the context of Special Relativity, but we have set $c = 1$ for convenience.

Appendix III

THE UNCOUNTABLE REAL NUMBERS

All we need to consider is the real numbers between 0 and 1. If they are uncountable, then so are the rest. Let us suppose that every real number between 0 and 1 can be put into a one-to-one correspondence with the positive integers, so that they have the same cardinality.

As we are assuming this can be done, there must be some scheme that associates each real number with each positive integer. Therefore, such a correspondence will associate some real decimal number between 0 and 1 with the number 1, another decimal number with 2, another with 3, and so forth. Let us write this down in a systematic way, as if we actually knew what the correspondences were (though of course we do not):

$$1 \leftrightarrow 0.\mathbf{2}8476298\ldots$$
$$2 \leftrightarrow 0.5\mathbf{6}962781\ldots$$
$$3 \leftrightarrow 0.84\mathbf{5}65083\ldots$$
$$4 \leftrightarrow 0.750\mathbf{0}0000\ldots$$
$$\vdots$$
$$n \leftrightarrow 0.24273961\ldots$$
$$\vdots$$

Now what we will do is create a real number, r, between 0 and 1, that cannot appear in our listing.

Our real number, r, will be of the form

$$r = 0.r_1 r_2 r_3 r_4 \ldots r_n \ldots$$

so let us choose the individual digits in a clever way, *à la* Cantor; namely, let us simply consider only the numbers in **bold** in the above scheme that are along the diagonal. The first digit is 2 and let us add 1 to get the number 3. This will be our first digit of *r*; that is, $r_1 = 3$. The next bold digit on the diagonal is 6, so we add 1 to get 7 and let $r_2 = 7$. Likewise, the next bold digit on the diagonal is 5, so that $r_3 = 5 + 1 = 6$.

Whenever we run into a 9 on the diagonal, we simple change it to 0 for the corresponding digit of *r*. In general,

$$r_n = nth\,digit\,on\,diagonal + 1$$

whenever the *nth* digit is less than 9, and 0 otherwise.

Continuing in this fashion what do we find? For one thing, our new number *r* differs from the first number in the scheme in at least the first decimal place. Moreover, the second digit of *r* differs from the second number in the scheme in at least the second decimal place. Likewise, the third digit of *r* differs from the third number in the list in the third decimal place. In general, the *nth* digit of *r* differs from the *nth* number in the list in the *nth* decimal place, and so on.

Can the number *r* be *somewhere* on our list? No it cannot, because it will differ from the first number in at least the first decimal place, from the second number in at least the second decimal place, from the third number in at least in the third decimal place, and so forth.

What all of this demonstrates is that there can be no enumeration of the real numbers between 0 and 1, and hence by extension all the real numbers, because there must always be real numbers left out from any such enumeration. This means that our assumption in the beginning could not possibly be true, and so it follows that there is no one-to-one correspondence between the positive integers and the real numbers.

The former are called *countable* for obvious reasons, and the real numbers are *uncountable*, as our proof, first demonstrated by Georg Cantor and known as the 'Cantor diagonal method' shows. In the language of the book, $\aleph_0 < c$.

We are also using a bit of logic here, in the form of the law of the excluded middle; namely, either a statement is true or its negation is true. We assumed that the statement: 'There is an enumeration of the real numbers' to be true, but this led to a contradiction. So, its negation, 'there is no enumeration of the real numbers' must therefore be true.

Appendix IV

$c^2 = c$: SQUARE AND LINE HAVE SAME CARDINALITY

We will present the essence of the argument in order to convince the reader of the validity of the assertion.[1]

Let us consider a square whose sides are all of 1 unit, as in Fig. A.3. Any point interior to the square, $p_1 = (x, y)$, has two coordinates, x and y, each of which can be represented by an infinite decimal expansion

$$x = 0.a_1 a_2 a_3 a_4 a_5 a_6 \cdots$$

$$y = 0.b_1 b_2 b_3 b_4 b_5 b_6 \cdots$$

Then we simply form the number p_1', given by interlacing the digits:

$$p_1' = 0.a_1 b_1 a_2 b_2 a_3 b_3 a_4 b_4 a_5 b_5 a_6 b_6 \cdots,$$

and so every point $p(x, y)$ inside the square can be uniquely paired with a number p' on the line segment $(0, 1)$.

Conversely, by taking any number p_2' on the line segment $(0, 1)$ given by

$$p_2' = 0.c_1 c_2 c_3 c_4 c_5 c_6 \cdots,$$

we can construct a unique number $p_2 = (x, y)$, paired with it in the square by employing the inverse of the above procedure to give the coordinates:

$$x = 0.c_1 c_3 c_5 \cdots$$

$$y = 0.c_2 c_4 c_6 \cdots$$

---

[1] For one thing, we will not concern ourselves with the details of decimal numbers having more than one representation, such as 0.50000... and 0.49999...

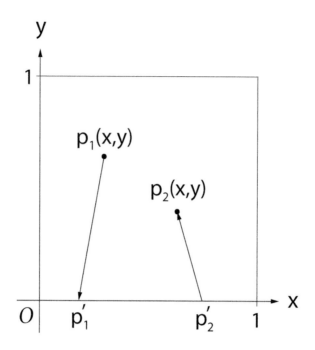

Figure A.3 There is a one-to-one correspondence between the interior points of the unit square and the points in the interval $(0, 1)$. (Illustration courtesy of Katy Metcalf.)

This illustrates how to establish a one-to-one correspondence between points p in the square with points p' on the line segment.[2] Thus, both sets have the same cardinality, i.e., $c^2 = c$. Indeed, $c^3 = c$, meaning there are not additional points to be found in a three-dimensional cube either.

---

[2] Cantor famously remarked, 'I see it, but I don't believe it'. The reader no doubt feels the same.

Appendix V

GEOMETRIC SERIES

Geometric series are mathematically very important and have the general form

$$1 + r + r^2 + r^3 + \cdots + r^n + \cdots$$

where r is some number between 0 and 1. That is, we wish to find the sum of an infinite series consisting of higher and higher powers of some number. In the example in the text

$$\frac{1}{2} + \frac{1}{4} + \frac{1}{8} + \frac{1}{16} + \cdots + \frac{1}{2^n} + \cdots$$

we have $r = 1/2$, and just note for later that the first term here begins with r instead of 1. We will take care of these matters at the end.

In order to determine the sum of the series, we need to consider the long-term behavior of the partial sums

$$s_n = 1 + r + r^2 + r^3 + \cdots + r^n.$$

To do this, let us now consider the product,

$$rs_n = r + r^2 + r^3 + r^4 \cdots + r^{n+1}.$$

Let us subtract this expression from the preceding one, which gives

$$s_n - rs_n = 1 - r^{n+1},$$

© Springer Nature Switzerland AG 2020
J. L. Schiff, *The Mathematical Universe*, Springer Praxis Books,
https://doi.org/10.1007/978-3-030-50649-0

as all the intermediate terms disappear, leaving only the first and last. The reason for doing this is that we can now obtain a nice expression for s_n by factoring the left-hand side,

$$s_n - rs_n = s_n(1-r) = 1 - r^{n+1},$$

and dividing both sides by $(1-r)$ yields:[1]

$$s_n = \frac{1 - r^{n+1}}{1-r}.$$

Since r is just some number between 0 and 1, it follows that higher and higher powers of r will just shrink to zero; that is, $r^{n+1} \to 0$ as $n \to \infty$. Therefore,

$$s_n \to \frac{1}{1-r},$$

as $n \to \infty$. This limit is a well-defined finite number and represents the sum of the geometric series we started with, namely,

$$1 + r + r^2 + r^3 + \cdots + r^n + \cdots = \frac{1}{1-r}.$$

For the geometric series in the text, we had $r = 1/2$. This would give,

$$1 + \left(\frac{1}{2}\right)^1 + \left(\frac{1}{2}\right)^2 + \left(\frac{1}{2}\right)^3 + \left(\frac{1}{2}\right)^4 + \cdots =$$

$$1 + \frac{1}{2} + \frac{1}{4} + \frac{1}{8} + \frac{1}{16} + \cdots = \frac{1}{1 - \frac{1}{2}} = 2.$$

All that is left to do is to subtract 1 from each side to give the original form of our geometric series, which now yields us the result:

$$\frac{1}{2} + \frac{1}{4} + \frac{1}{8} + \frac{1}{16} + \cdots = 1.$$

There You Have It. Note that by substituting $-r$ for r into the geometric series we obtain the alternating geometric series and its sum,

$$1 - r + r^2 - r^3 + r^4 - r^5 + \cdots = \frac{1}{1+r}.$$

---

[1] For the geometric series in the text, we had $r = 1/2$, so that the expression for the partial sums would read: $s_n = \dfrac{1 - \left(\dfrac{1}{2}\right)^{n+1}}{1 - 1/2} = 2\left(1 - \dfrac{1}{2^{n+1}}\right) = 2 - \dfrac{1}{2^n}$. As our geometric series in the text did not start with the value 1, we must subtract 1 from the expression for s_n, which gives: $s_n = 1 - \dfrac{1}{2^n}$, as mentioned in Footnote 15 of Chapter 2.

Appendix VI

CESÀRO SUMS

If $s_1, s_2, s_3, \ldots$, denote the partial sums of an infinite series as given in Chapter 2, we now take the averages of these sums,

$$\sigma_1 = s_1; \; \sigma_2 = \frac{s_1 + s_2}{2}; \; \ldots; \; \sigma_n = \frac{s_1 + s_2 + \ldots + s_n}{n}.$$

Computing these averages from the Grandi series

$$1 - 1 + 1 - 1 + 1 - \cdots$$

we find from the partial sum values that

$$\sigma_1 = 1, \; \sigma_2 = \frac{1}{2}, \; \sigma_3 = \frac{2}{3}, \; \sigma_4 = \frac{2}{4}, \; \sigma_5 = \frac{3}{5}; \; \sigma_6 = \frac{3}{6}; \; \ldots.$$

These σ_n can be seen to converge to the value 1/2 as n increases, since one can write the values of σ_n in the form

$$\sigma_n = \frac{1}{2} + \frac{1}{2n}, \text{ whenever } n > 1 \text{ is odd,}$$

and

$$\sigma_n = \frac{1}{2}, \text{ when } n \text{ is even.}$$

As n becomes arbitrarily large, $\frac{1}{2n}$ approaches zero, and so σ_n is approaching the value of 1/2, which represents the Cesàro sum of the Grandi series

$$1 - 1 + 1 - 1 + 1 - \ldots = 1/2.$$

© Springer Nature Switzerland AG 2020
J. L. Schiff, *The Mathematical Universe*, Springer Praxis Books,
https://doi.org/10.1007/978-3-030-50649-0

Appendix VII

ROTATING A VECTOR VIA A QUATERNION

Suppose one wishes to rotate a vector **v** in three-dimensional space, through an angle θ about an axis given by the unit vector[1]

$$\mathbf{u} = (x, y, z).$$

One then forms the quaternion, q, given by

$$q = w + q_1 i + q_2 j + q_3 k,$$

where

$$w = \cos\left(\frac{\theta}{2}\right), \; q_1 = x\sin\left(\frac{\theta}{2}\right), \; q_2 = y\sin\left(\frac{\theta}{2}\right), \; q_3 = z\sin\left(\frac{\theta}{2}\right),$$

and the values of x, y, z are the coordinates of the vector **u**. Thus, we have incorporated the axis of rotation **u** and the angle of rotation θ into our quaternion.

---

[1]A unit vector is a vector that has length one. Any (non-zero) vector can be made into a unit vector by dividing the vector by its own length. The same direction in space will be preserved as the original vector. The length of **u**, $\|\mathbf{u}\|$, in three-space is given by:

$$\|\mathbf{u}\| = \left(x^2 + y^2 + z^2\right)^{1/2},$$

as was discussed in Chapter 7 (Infinite Space).

© Springer Nature Switzerland AG 2020
J. L. Schiff, *The Mathematical Universe*, Springer Praxis Books,
https://doi.org/10.1007/978-3-030-50649-0

By performing the quaternion multiplication

$$\mathbf{v}' = q\,\mathbf{v}\,q^{-1},$$

the vector $\mathbf{v}'$ represents the rotated vector $\mathbf{v}$ through the desired angle θ.

For example, suppose we would like to rotate the vector $\mathbf{v} = (2,2,0)$ through $180°$ about the axis determined by the unit vector $\mathbf{u} = (1,0,0)$ as in Fig. A.4.

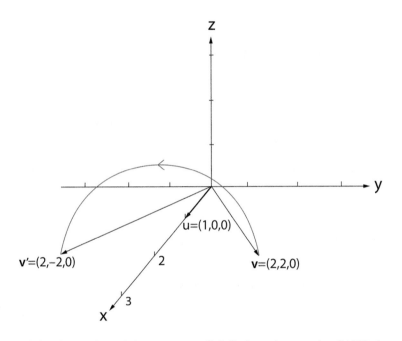

Figure A.4 The rotation of the vector $\mathbf{v} = (2,2,0)$ through an angle of $180°$ about the axis determined by the unit vector $\mathbf{u} = (1,0,0)$, i.e. the x-axis. (Illustration courtesy of Katy Metcalf.)

Therefore: $w = cos\left(\dfrac{\theta}{2}\right) = 0,$ and $sin\left(\dfrac{\theta}{2}\right) = 1,$ so   that   our   quaternion $q = w + q_1 i + q_2 j + q_3 k$, takes the form

$$q = 0 + i + 0j + 0k.$$

This simple example was chosen to keep the computation to a minimum, so that the author could do it and so that we would already know the answer. It turns out that since q is already a unit vector, the inverse is given by: $q^{-1} = (0 - i - 0j - 0k).^2$ Therefore, the rotated vector is:

$$\mathbf{v}' = q\mathbf{v}q^{-1}$$

$$= (0 + i + 0j + 0k)(0 + 2i + 2j + 0k)(0 - i - 0j - 0k).$$

We need to be careful of the order since the multiplication does not commute. Multiplying first:

$$\mathbf{v}q^{-1} = (0 + 2i + 2j + 0k)(0 - i - 0j - 0k) = (2 + 0i + 0j + 2k).$$

Then we multiply by q to obtain:

$$\mathbf{v}' = q\mathbf{v}q^{-1} = (0 + i + 0j + 0k)(2 + 0i + 0j + 2k)$$

$$= (0 + 2i - 2j + 0k).$$

This is just what we expected, namely that the rotated vector is $(2, -2, 0)$.

---

[2] In general, the inverse q^{-1} is defined to be:

$$q^{-1} = \frac{1}{w^2 + q_1^2 + q_2^2 + q_3^2}(w - q_1 i - q_2 j - q_3 k).$$

Now in our case, for the expression in the denominator,

$$w^2 + q_1^2 + q_2^2 + q_3^2 = \left(\cos^2(\theta/2) + (x^2 + y^2 + z^2)\sin^2(\theta/2)\right)$$

$$= 1,$$

since $\|\mathbf{u}\| = (x^2 + y^2 + z^2)^{1/2} = 1$ and since $\sin^2(\varphi) + \cos^2(\varphi) = 1$ from trigonometry. That leaves

$$q^{-1} = w - q_1 i - q_2 j - q_3 k.$$

Appendix VIII

QUATERNIONS: $q^2 = -1$

In order to find all the quaternion solutions to the equation $q^2 = -1$, we let $q = w + xi + yj + zk$. Therefore, utilizing the rules for multiplication of i, j, k, we have, by brute force calculation:

$$-1 = q^2 = (w + xi + yj + zk) \times (w + xi + yj + zk)$$

$$= (w^2 - x^2 - y^2 - z^2) + (wx + xw + yz - zy)i +$$
$$(wy + yw - xz + zx)j + (wz + zw + xy - yx)k$$

$$= (w^2 - x^2 - y^2 - z^2) + (2wx)i + (2wy)j + (2wz)k.$$

Now, if two quaternions are equal, then their corresponding coordinates must be identical on the left and right side of the equation, which means that

$$(w^2 - x^2 - y^2 - z^2) = -1, \qquad 2wx = 2wy = 2wz = 0.$$

From the three second equalities, if $w \neq 0$, that would mean that $x = y = z = 0$, resulting in, $w^2 = -1$, which is not possible as w is a *real* number. It follows that $w = 0$, and from the first expression we obtain

$$x^2 + y^2 + z^2 = 1.$$

This equation represents, in three-dimensional space, all the points (x, y, z) on the sphere of radius $= 1$, centered at the origin. For any of these points, we have $q^2 = -1$, where our (pure) quaternion is given by: $q = xi + yj + zk$. Thus, there are *infinitely many* quaternions q satisfying: $q = \sqrt{-1}$, unlike for the complex numbers where there are only two: $z = \pm i$.

© Springer Nature Switzerland AG 2020
J. L. Schiff, *The Mathematical Universe*, Springer Praxis Books,
https://doi.org/10.1007/978-3-030-50649-0

269

Appendix IX

RIEMANN ZETA FUNCTION

For those who absolutely must know what the Riemann zeta function looks like in its entirety, defined in the whole complex plane, here it is. But do not be daunted by its appearance.

Firstly, we must extend the zeta function of eq. (13), which was only defined in the region $Re(s) > 1$ to the 'critical strip' $0 < Re\,(s) < 1$. We do this by considering the *alternating* zeta function which converges for $Re(s) > 0$

$$\eta(s) = \sum_{n=1}^{\infty} \frac{(-1)^{n-1}}{n^s} = \left(1 - 2^{1-s}\right)\zeta(s).$$

Solving for $\zeta(s)$ gives the expression

$$\zeta(s) = \left(\frac{1}{1-2^{1-s}}\right)\sum_{n=1}^{\infty}\frac{(-1)^{n-1}}{n^s},$$

which now defines the zeta function for all of $Re(s) > 0$. So far, so good.

The final formulation for the Riemann zeta function is actually more harmless than it looks, namely:

$$\zeta(s) = 2^s\,\pi^{s-1}\,sin\left(\frac{\pi s}{2}\right)\Gamma(1-s)\zeta(1-s). \tag{35}$$

where s is any complex number. The above equation demonstrates how the zeta function takes values in the left-half plane, $Re(s) < 0$, directly from the value at its reflected counterpart $1 - s$ in the right-half plane $Re(s) > 1$.

© Springer Nature Switzerland AG 2020
J. L. Schiff, *The Mathematical Universe*, Springer Praxis Books,
https://doi.org/10.1007/978-3-030-50649-0

Let us take the function apart, bit by bit, and see just how harmless it really is. Reading eq. (35) from left to right, we have the number 2 raised to the power of our complex variable s. Next is the number π raised to the power $s - 1$. Then comes the *sine* of the complex number $(\pi s/2)$. This presents no problem, as all the trigonometric functions can be extended to complex numbers. In the case of $sin(z)$, this is given by the infinite series of eq. (6), replacing the x by z. Thus, $sinx$ and $sinz$ completely agree when z is a real number. So far, so good, and not scary at all.

Next is a new function, $\Gamma(s)$, called the *gamma function*, which for real numbers is just an extension of the notion of factorials.[1] The final ingredient is the appearance of the zeta function again, but here defined on the number $(1 - s)$. Admittedly, this looks a bit strange, but let us just go with it and see what happens.

From a very elementary perspective, can we say anything interesting about this strange looking Riemann zeta function? Yes we can. Note that when $s = -2, -4, -6, \ldots$, that is, at any of the negative even numbers, one component in eq. (35) is,

$$sin\left(\frac{\pi s}{2}\right) = 0,$$

so that, $\zeta(s) = 0$, and these are the well-known *trivial zeros* of the zeta function, as discussed in the text.

Of course, $sin\left(\frac{\pi s}{2}\right) = 0$, whenever s is also a positive even integer, *but*, the quantity next to it in eq. (35), namely: $\Gamma(1 - s)$, becomes infinite as the number s approaches any positive even integer. This approach to infinity happens to be in such a way as to cancel out the *sine* term which is approaching zero in value. The result of their product works out to be a non-zero value at every positive even integer, s. Indeed, just as in eq. (5), we will get:

$$\frac{\pi^2}{6} = \zeta(2) = \frac{1}{1^2} + \frac{1}{2^2} + \frac{1}{3^2} + \frac{1}{4^2} + \frac{1}{5^2} + \cdots,$$

and likewise, for $\zeta(4) = \frac{\pi^4}{90}$, as noted in the discussion on the zeta function in the text.

At this stage, let us be daring enough to put the value $s = -1$ into the Riemann zeta function given by eq. (35), like so:

$$\zeta(-1) = 2^{-1} \pi^{-2} sin\left(\frac{-\pi}{2}\right) \Gamma(2) \zeta(2).$$

---

[1] For positive integers, n, the gamma function takes the values: $\Gamma(n) = (n - 1)!$, and in between the positive integers the function has interpolated values that lie on a continuous curve. The function can also be extended in a natural way to negative numbers as well as to complex numbers.

Since, $sin\left(\dfrac{-\pi}{2}\right) = -1$, $\Gamma(2) = 1! = 1$, and $\zeta(2) = \pi^2/6$, we find by putting in these corresponding values that,

$$\zeta(-1) = \frac{1}{2}\frac{1}{\pi^2}(-1)(1)\left(\frac{\pi^2}{6}\right) = -1/12,$$

as mentioned in the text.

To get some appreciation of the behavior of the Riemann zeta function $\zeta(s)$, the best way is to graph the absolute value, $|\zeta(s)|$. This is always a real number and gives the magnitude of $\zeta(s)$ at each value of s which can be plotted on a third (elevation) axis, as in Fig. A.5. The peak in the middle is the pole at $s = 1$ that actually goes up to an infinite height. The sharp points where the function is dipping down to the complex plane are the zeros on the critical line $x = 1/2$.

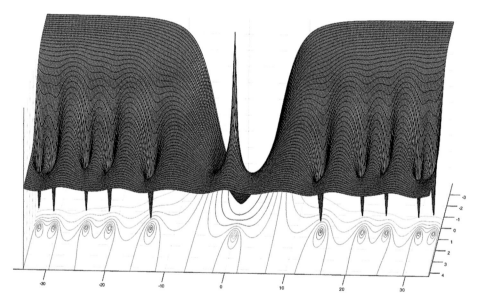

Figure A.5 A plot of a small region for the Riemann zeta function. The vertical axis represents $|\zeta(s)|$ for $-3 \le x \le 4$, and, $-30 \le y \le 30$. The imaginary axis runs horizontally. One can see how the function dips right down to zero at various points on the critical line $x = \dfrac{1}{2}$ which are the positions of the nontrivial zeros mentioned in the text. The sharp upward peak is at the point $s = 1$ where the function $\zeta(s)$ has a pole and becomes infinite. Contour maps of what is happening above the complex plane lie in the forefront of the image in light blue. (Image courtesy of Chris King.)

Lastly, let us mention the immediate connection between the Riemann zeta function and all of the prime numbers. It is a wonderful equality known as the Euler product formula:

$$\zeta(s) = \left(\frac{1}{1-2^{-s}}\right)\left(\frac{1}{1-3^{-s}}\right)\left(\frac{1}{1-5^{-s}}\right)\left(\frac{1}{1-7^{-s}}\right)\cdots\left(\frac{1}{1-p^{-s}}\right)\cdots,$$

with the infinite product being taken over all the primes 2, 3, 5, 7, 11, 13, 17, …, in the denominators, and s is any complex number whose real part is greater than 1.[2]

From this product expression, taking $s = 2$, and the fact that $\zeta(2) = \pi^2/6$, we may conclude that

$$\frac{\pi^2}{6} = \left(\frac{1}{1-2^{-2}}\right)\left(\frac{1}{1-3^{-2}}\right)\left(\frac{1}{1-5^{-2}}\right)\left(\frac{1}{1-7^{-2}}\right)\cdots$$

$$= \frac{4\cdot9\cdot25\cdot49\cdot121\cdots}{3\cdot8\cdot24\cdot48\cdot120\cdots},$$

which is another remarkable relation involving our beloved π, with an infinite product comprising all the squares of primes. Almost too good to be true.[3]

---

[2] Euler's original proof of the product formula was for the variable s being a real number with $s > 1$. Bear in mind that his work on the product formula preceded that of Riemann's by more than 100 years. Moreover, the product formula remains valid for any complex number s whose real part is greater than 1, where the Riemann zeta function is given by the expression in eq. (13).

[3] Additionally, we have already seen an infinite sum representation for $\pi^2/6$ given by the p-series for $p = 2$, which is the same as $\zeta(2)$.

Appendix X

RANDOM WALK CODE

Python code for a random walk simulation, starting at a point inside a circle (of size of the user's choosing) with $100°$ on the top half and $0°$ on the bottom half. Selecting $10,000$ steps and a step size of 0.1 gives a reasonably accurate determination of the actual temperature at the starting point. A Python compiler is freely available online at: https://repl.it/languages/Python3. Code courtesy of Ryan Schiff:

© Springer Nature Switzerland AG 2020   274
J. L. Schiff, *The Mathematical Universe*, Springer Praxis Books,
https://doi.org/10.1007/978-3-030-50649-0

```python
import operator
import math
from math import pi, cos, sin
from random import random
from decimal import Decimal as D
operators=[('+',operator.add),('-',operator.sub)]
x = D(input("Enter X Coordinate:"))
y = D(input("Enter Y Coordinate:"))
r = D(input("Radius of circle:"))
d = D(input("Distance of step:"))
n = int(input("Enter number of cycles:"))
count = 0
for i in range(n):
    theta = random() * 2 * pi
    a = x + D(cos(theta)) * d
    b = y + D(sin(theta)) * d
    B = math.sqrt(D((a**2))+ D((b**2)))
    if D(B)>=r:
        count+=1

    else:
        while B<r:
            theta = random() * 2 * pi
            a = a + D(cos(theta)) * d
            b = b + D(sin(theta)) * d
            B = math.sqrt(D((a**2))+ D((b**2)))
            if D(B)>=r and D(b)>0:
                count+=1


i=i+1

prob = (count)/n
degrees = (prob) * 100
print (degrees)
```

Appendix XI

AGE OF THE SOLAR SYSTEM

The calculations below are nothing more than high school algebra, but they look a bit daunting because we are obliged to use the correct scientific notation for the isotopes involved. However, there is nothing to fear as it is all based on the formula for exponential decay.

The radioisotope of rubidium we employ uses the exponential formula of eq. (14) in the text that we applied to rubidium-strontium decay:

$$^{87}Rb = {}^{87}Rb_0 e^{-kt}, \tag{36}$$

and we know that rubidium-87 decays into strontium-87 with a decay constant given by $k = 1.42 \times 10^{-11}$. In eq. (36), $^{87}Rb_0$ represents the quantity of rubidium that was present initially when it became part of the meteorite, and ^{87}Rb represents the amount remaining after the elapsed time t. This amount is decreasing with time, as given by the formula, indicating exponential decay.

Let $^{87}Sr_0$ represent the initial amount of strontium-87 in the meteorite sample, which was there naturally from the beginning. Although it is a quantity we cannot possibly know the value of, we can still carry on regardless and write the equation:

$$^{87}Sr = {}^{87}Sr_0 + \left({}^{87}Rb_0 - {}^{87}Rb \right). \tag{37}$$

This is just the mathematical way to express that the amount of strontium-87 in the meteorite *now*, namely, ^{87}Sr, is equal to the amount that was there initially from the start, $^{87}Sr_0$, plus the amount contributed by the decay of rubidium-87 into strontium-87. This latter is given by the quantity:

$$^{87}Rb_0 - {}^{87}Rb,$$

on the far-right side of eq. (37).

© Springer Nature Switzerland AG 2020
J. L. Schiff, *The Mathematical Universe*, Springer Praxis Books,
https://doi.org/10.1007/978-3-030-50649-0

So far, so good. Of course, there are now *two* quantities that we do not know, which are the initial amounts of strontium and rubidium — $^{87}Sr_0$, and $^{87}Rb_0$, respectively — because no one was around to measure them. Nevertheless, we can eliminate one of the unknown quantities in eq. (37) by substituting into this equation the expression for $^{87}Rb_0$ obtained from our initial eq. (36). Thus eq. (37) becomes, upon simplifying,

$$^{87}Sr = \ ^{87}Sr_0 + \left(\ ^{87}Rb \ e^{kt} - \ ^{87}Rb \right) = \ ^{87}Sr_0 + \ ^{87}Rb \left(e^{kt} - 1 \right).$$

Next comes the clever part, since there still remains the unknown initial quantity of strontium-87 in this last formulation. We *divide* the terms of the left and right sides of this last equation by the quantity of strontium-86, which is neither radioactive nor produced via radioactive decay and thus remains constant over time. Why we do this will become apparent shortly.

This gives, after rearranging the terms to look more familiar,

$$\frac{^{87}Sr}{^{86}Sr} = \left(e^{kt} - 1 \right) \frac{^{87}Rb}{^{86}Sr} + \frac{^{87}Sr_0}{^{86}Sr}. \tag{38}$$

Now, eq. (38) looks even more complicated and it appears that we have made things worse for ourselves, but notice that it has the form of the formula for a straight line: $y = mx + b$, where

$$y = \frac{^{87}Sr}{^{86}Sr}; \quad m = \left(e^{kt} - 1 \right); \quad x = \frac{^{87}Rb}{^{86}Sr}; \quad \text{and } b = \frac{^{87}Sr_0}{^{86}Sr}.$$

Both of the ratio quantities: $x = \dfrac{^{87}Rb}{^{86}Sr}$, and $y = \dfrac{^{87}Sr}{^{86}Sr}$, are scientifically *measurable* from the meteorite using an instrument called a mass spectrometer. Taking several samples, their values (x, y) can be plotted on a graph with an x-axis and y-axis, and then we can draw a best fit straight line through the data points. That is the reason for dividing everything by the quantity ^{86}Sr. This procedure neatly eliminates having to deal with the unknown quantity $^{87}Sr_0$, since we can draw our straight line without it.

Once the straight line is drawn, the slope m of the line can be measured by, $m = \dfrac{\Delta y}{\Delta x}$. Since $m = (e^{kt} - 1)$, and we know the value of the decay constant, k, the elapsed time t is easily determined.

In Fig. A.6 is a graph of actual rubidium/strontium data, with the slope measured at $m = 0.3/4.51 = 0.0665$.

Using

$$\left(e^{kt} - 1 \right) = m = 0.0665,$$

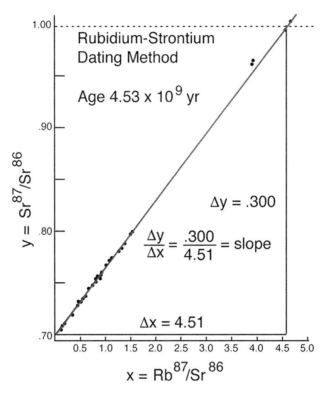

Figure A.6 A rubidium/strontium straight line graph from meteorite samples, used to determine the age of the Solar System. (From G.W. Wetherill, *Ann. Rev. Nucl. Sci.* 25, 283, 1975.)

with a decay constant value: $k = 1.42 \times 10^{-11}$,[1] and then solving for t, gives

$$t = \left(\frac{1}{k}\right) \ln(m+1) = \left(0.704 \times 10^{11}\right) \ln(1.0665) = 4.53 \times 10^{9} \; years.$$

Here, ln is the natural logarithm base e and our value agrees with that of Wetherill (above). Of course, some samples from other meteorites have proven slightly older than this one, with the oldest age being measured at 4.5689 billion years, when the first material started to solidify out of the swirling cloud of gas and dust known as the Solar Nebula. This is the age of the Solar System.

---

[1] The value of the decay constant k for rubidium-87 has been refined somewhat in recent years giving a value of: $k = 1.398 \times 10^{-11}$.

Appendix XII

CHELYABINSK METEOROID

Let us compute the energy of the Chelyabinsk asteroid event. It had an estimated diameter of $17 - 20$ meters, so let us take 18 m. First, we need the mass of the object, and we will assume it was spherical so that we can use the standard equation for the volume of a sphere,

$$V = \frac{4}{3}\pi r^3,$$

where in our case, $r = 9$ m. Doing the calculation gives a volume equal to 3,054 m^3. The meteoroid is known to be stony, so let us take an average density for a stony meteorite of 3 $grams/cm^3$, which equates to 3, 000 kg/m^3.

In order to get the mass of the meteoroid, we multiply *mass* × *density*, yielding a mass of 9.16×10^6 kg. That is 9, 000 metric tons by the way.

Now we use the equation for kinetic energy

$$E = \frac{1}{2}mv^2,$$

where the velocity has been given as: $v = 19\frac{km}{sec} = 19,000$ m/sec. Therefore, the energy is

$$E = 0.5\left(9.16 \times 10^6\right)\left(1.9 \times 10^4\right)^2 \approx 16.5 \times 10^{14}\ joules,$$

But we would like to convert this to something more meaningful, like the energy equivalent of TNT. Since 1 kiloton of TNT equals 4.185×10^{12} *joules*, dividing E by this factor gives the energy equivalent of roughly 400 kilotons. For comparison, the first atomic bomb tested in Alamogordo, New Mexico, had a yield of about 20 kilotons of TNT.

These calculations show that nuclear weapons are not the only threat from the sky, and that large asteroid fragments are potentially even more hazardous.

© Springer Nature Switzerland AG 2020
J. L. Schiff, *The Mathematical Universe*, Springer Praxis Books,
https://doi.org/10.1007/978-3-030-50649-0

Appendix XIII

LOGIC GATES

The three main logic gates derive from logic and are a part of Boolean Algebra, named for English mathematician/logician George Boole (1815–1864) who did seminal work in this area. Let us start by using tables (*truth tables*), to determine the truth or falsity of a compound statement based on the truth or falsity of their component statements.

Examples of a compound statement are:

She went to the store **and** she bought a loaf of bread.
You can pay me now **or** you can pay me later.

We will consider the following three main cases for True = T, False = F, where all possibilities of T and F are considered for statements P and Q:

P	Q	P AND Q	P	Q	P OR Q	P	NOT P
F	F	F	F	F	F	T	F
F	T	F	F	T	T	F	T
T	F	F	T	F	T		
T	T	T	T	T	T		

The left table represents a conjunction (and), the center table a disjunction (or) and the right table a negation (not).

If we now think of P and Q as electrical inputs A and B that can either be 1 = T or 0 = F, the above truth tables become:

© Springer Nature Switzerland AG 2020
J. L. Schiff, *The Mathematical Universe*, Springer Praxis Books,
https://doi.org/10.1007/978-3-030-50649-0

A	B	A AND B	A	B	A OR B	A	NOT A
0	0	0	0	0	0	1	0
0	1	0	0	1	1	0	1
1	0	0	1	0	1		
1	1	1	1	1	1		

Interpreting, as we have in Chapter 5, a switch being ON with 1 and OFF with 0, the above tables defines the output of an electronic AND, OR, and NOT gate, respectively. Therefore, the AND gate would have two inputs, A and B, each being either 1 or 0:

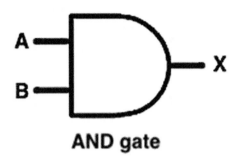

AND gate

and its output, X, would be in accordance with the preceding AND table. Similarly, the OR gate has two inputs, A and B, each being 1 or 0:

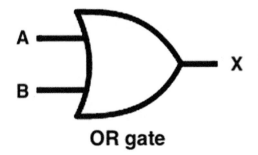

OR gate

with the output again given by X in accordance with the preceding OR table. Finally, the NOT gate (known as an *inverter* in electronics) has a single input, A, being 1 or 0:

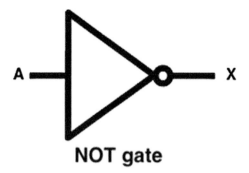

NOT gate

whose output, X, is governed by the preceding NOT table. The above represent the standard symbols used for each gate.

By combining these three basic logic gates, it is possible to construct many complex electronic circuits which form the basis for a general-purpose computer.

Appendix XIV

GALAXY DISTANCE VIA CEPHEIDS

Let us compute the distance to the Large Magellanic Cloud (LMC, Fig. A.7), one of our nearest neighboring galaxies and the one studied by Henrietta Leavitt in determining the relationship between the period and intrinsic brightness of Cepheid variables.

This relationship can be conveyed by a mathematical expression in the form of a straight line:

$$M = -2.43 \log_{10} P - 1.62. \tag{39}$$

Here, the period P of the Cepheid is measured in days and the quantity M is a measure of the *absolute magnitude*.[1] The technical meaning of this latter term is the (apparent) brightness of a star when it is viewed from a standard distance of 10 parsecs (32.6 light-years). In this way, the intrinsic brightness of one star can be compared to that of any another in a standardized fashion. Just place them (theoretically of course) at the same distance of 10 parsecs and observe how bright they appear.

We just need one other formula that relates the absolute magnitude M to the *apparent magnitude m*, the quantity that astronomers actually measure through their telescopes. Of course, this magnitude is going to involve the distance d to the object, namely,

---

[1] Formula from: G.F. Benedict *et al.*, Hubble Space Telescope fine guidance sensor parallaxes of galactic Cepheid variable stars: period-luminosity relations, *Astrophys. J.*, Vol. 133, pp. 1810–1827, 2007. Note that some stars have an absolute magnitude that is negative in value. For example, the star Rigel in the Orion nebula has an absolute magnitude of −7.84.

© Springer Nature Switzerland AG 2020
J. L. Schiff, *The Mathematical Universe*, Springer Praxis Books,
https://doi.org/10.1007/978-3-030-50649-0

$$m - M = 5\log_{10} d - 5. \tag{40}$$

Once we measure the period P of a Cepheid variable, we put it into eq. (39) which then yields the absolute magnitude M. Coupled with the observed apparent magnitude m, we can readily solve for the distance d in eq. (40).

Let us take some real data and do the computation. One particular Cepheid, catalogued as HV886, has a period of 24 days.[2] Therefore, $\log_{10}P = 1.38$. Putting this value into eq. (39) gives the absolute magnitude value:

$$M = -2.43(1.38) - 1.62 = -4.97.$$

The measured apparent magnitude for this star is $m = 13.47$. Hence, by eq. (40),

$$m - M = 13.47 - (-4.97) = 18.44$$

$$= 5\log_{10} d - 5.$$

Rearranging terms,

$$\log_{10}d = \frac{18.44 + 5}{5} = 4.69.$$

Then the distance is,

$$d = 10^{4.69} \approx 49{,}000 \text{ parsecs} \approx 160{,}000 \text{ light-years}.$$

Of course, in practice, numerous Cepheids are observed in any given situation and the average distance is computed. Our value is very close to a recent distance determination for the LMC, by other means, of 49.59 kiloparsecs $\approx$ 161,660 light-years.[3]

Both the Large and Small Magellanic Clouds no doubt beguiled our African ancestors in the very distant past.

---

[2] Data from P. Karczmarek *et al.*, Large Magellanic Cloud Cepheids in the ASAS data, *Acta Astrometrica,.*, Vol. 61, pp. 303–318, 2011.

[3] G. Pietrzyński *et al.*, A distance to the Large Magellanic Cloud that is precise to one percent, *Nature*, Vol. 567, pp. 200–203, 2019. The observations used eclipsing binary stars.

Figure A.7 One of our nearest neighboring galaxies, the Large Magellanic Cloud that is approximately 162,000 light years distant and easily visible with the naked eye from the Southern Hemisphere. (Image courtesy of NASA.)

Appendix XV

TIME DILATION

From our discussion in the Time in Motion section of Chapter 7, we determined the time duration of a single tick of our stationary clock and that of the clock aboard a rocket ship.

In our first scenario, the light source and the mirror are stationary. So how much time does it take for one tick of our light clock? We know from everyday experience that if you are travelling at 100 kilometers per hour in your car for a period of 3 hours, then you have covered $100 \times 3 = 300$ kilometers. In other words,

$$distance = velocity \times time,$$

which can be written symbolically as

$$d = vt.$$

In our clock setting, the distance equals $2d$ (from A to B and back), and so solving for the time duration t, given that the velocity of light is the constant c, then

$$2d = ct,$$

or, solving for t,

$$t = \frac{2d}{c},$$

which is the length of time of a single tick of our stationary clock. Let us call this time t_0 to indicate that it represents the time duration of a tick of the *stationary* clock, so that

$$t_0 = \frac{2d}{c}.$$

© Springer Nature Switzerland AG 2020
J. L. Schiff, *The Mathematical Universe*, Springer Praxis Books,
https://doi.org/10.1007/978-3-030-50649-0

For the clock aboard the moving rocket ship, we have the same relationship for the duration of a single tick, except that the distance travelled by the light beam in the clock is $2D$, so that

$$t = \frac{2D}{c}.$$

Solving for d in the former expression, we have $d = ct_0/2$, and solving for D in the latter expression gives $D = ct/2$. These are two of the components we require in our right triangle of Fig. A.8.

The third ingredient we need is the length of the base a in the triangle. Here, we use the fact that *distance = rate × time*. As the rocket ship is moving at a constant velocity $= v$, the distance a that it travels in half a tick of the moving clock is

$$a = v\left(\frac{t}{2}\right).$$

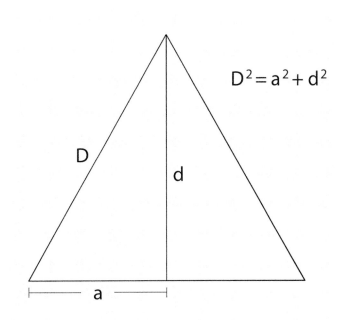

$$D^2 = a^2 + d^2$$

Figure A.8 The relation between the distance d for half a tick of a stationary light clock and the distance D of half a tick of a light clock that is moving relative to a stationary one. (Illustration courtesy of Katy Metcalf.)

Now we are in a position to relate the two distances d and D. Via Fig. A.8, we invoke the theorem of Pythagoras, the famed Greek philosopher and mathematician who lived 2,500 years ago. Since

$$D^2 = a^2 + d^2,$$

we can substitute the two values for a and d that we just determined above:

$$D^2 = a^2 + d^2 = v^2 \left(\frac{t}{2}\right)^2 + \left(\frac{ct_0}{2}\right)^2 = \frac{v^2 t^2}{4} + \frac{c^2 t_0^2}{4}. \tag{41}$$

Furthermore, since $D = ct/2$, we can also eliminate D in eq. (41), so that

$$D^2 = \frac{c^2 t^2}{4} = \frac{v^2 t^2}{4} + \frac{c^2 t_0^2}{4}.$$

Dispensing with the 4s and rearranging the terms, we obtain

$$t^2 \left(c^2 - v^2\right) = c^2 t_0^2,$$

and dividing both sides by c^2 yields

$$t^2 \left(1 - \frac{v^2}{c^2}\right) = t_0^2.$$

Thus, we have found that the time duration t of a tick of a moving clock with a constant velocity v for an observer on the ground is given by the equation,

$$t = \frac{t_0}{\sqrt{1 - \frac{v^2}{c^2}}}.$$

This is the formula for how time t slows down for a moving object, with respect to the time t_0 of a stationary observer noted in eq. (31). Thus, whenever $v > 0$, we have

$$\sqrt{1 - v^2/c^2} < 1,$$

as discussed in the text. As a consequence, $t > t_0$ which means that the tick of the clock aboard the rocket ship is of longer duration, and time passes more slowly relative to the passage of time on the ground.

Appendix XVI

EXPANSION OF THE UNIVERSE

To describe the expansion of the Universe over time, it is convenient to introduce what is called a scaling factor, $a(t)$. This just indicates how space expands as a function of time. One way to express how this works in terms of, say, the distance between two galaxies, is to let $D_0 = D(t_0)$ be the distance at some given moment in time t_0. As the Universe expands, the distance $D(t)$ at any future time t will be governed by the scaling factor:

$$D(t) = a(t)D_0, \tag{42}$$

or equivalently, $a(t) = D(t)/D(t_0)$.

The rate of change of $D(t)$ is just the radial velocity v, so that taking the derivative of $D(t)$, in eq. (42) we obtain

$$v = \frac{d\,D(t)}{dt} = \dot{a}(t)D_0,$$

where $\dot{a}(t)$ denotes the derivative of $a(t)$ with respect to t. Substituting the expression for D_0 from eq. (42) into the preceding equation yields the expression

$$v = \frac{\dot{a}(t)}{a(t)}D(t).$$

This is starting to look familiar, for if we set

$$H_t = \frac{\dot{a}(t)}{a(t)}, \tag{43}$$

J. L. Schiff, *The Mathematical Universe*, Springer Praxis Books,
https://doi.org/10.1007/978-3-030-50649-0

we have just derived the Hubble formula for the expansion of the Universe and done some real Cosmology. That is,

$$v = H_t D(t),$$

where the Hubble parameter H_t is defined as in eq. (43). This is the Hubble formula from eq. (22), where the time there represents the present epoch.

Bibliography

Aaronson, S.
Quantum Computing Since Democritus, Cambridge University Press, 2013.

Barrow, J.D. and Tipler, F.J.
The Anthropic Cosmological Principle, Oxford University Press, 1998.

Becker, A.
What is Real, Basic Books, 2018.

Beckmann, P.
A History of Pi, St. Martin's Press, 1976.

Bodanis, D.
$E = mc^2$: *A Biography of the World's Most Famous Equation*, Macmillan, 2000.

Butterworth, B.
The Mathematical Brain, Macmillan, 1999.

Carroll, S.
Something Deeply Hidden: Quantum Worlds and the Emergence of Spacetime, Dutton, 2019.

Davies, P.
The Goldilocks Enigma: Why Is the Universe Just Right for Life?, Mariner Books, 2008.

Derbyshire, J.
Prime Obsession: Bernhard Riemann and the Greatest Unsolved Problem in Mathematics, John Henry Press, 2003.

Devaney, R.L.
An Introduction to Chaotic Dynamical Systems, Addison-Wesley, 1989.

Epstein, R.L.
An Introduction to Formal Logic, Advanced Reasoning Forum, 2016.

Flake, G.
The Computational Beauty of Nature, The MIT Press, 2001.

© Springer Nature Switzerland AG 2020
J. L. Schiff, *The Mathematical Universe*, Springer Praxis Books,
https://doi.org/10.1007/978-3-030-50649-0

Gleick, J.
Chaos: The Making of a New Science, Penguin Books, 2008.

Greene, B.
The Elegant Universe, W.W. Norton, 2003.

Greene, B.
The Hidden Reality – Parallel Universes and the Deep Laws of the Cosmos, A. Knopf, 2011.

Gribbin, J.
Schödinger's Kittens, Phoenix, 1995.

Hardy, G.H.
A Mathematician's Apology, Cambridge University Press, 1967.

Hawking, S. and Mlodinow, L.
The Grand Design: New Answers to the Ultimate Questions of Life, Bantam Press, 2010.

Hofstadter, D.
Gödel, Escher, Bach: An Eternal Golden Braid, Basic Books, 1999.

Kauffman, S.A.
At Home in the Universe: The Search for Laws of Self-Organization and Complexity, Oxford University Press, 1995.

Lennox, J.C.
God's Undertaker – Has Science Buried God, Lion, 2009.

Livio, M.
Is God a Mathematician, Simon & Schuster, 2009.

Lloyd, S.
Programming the Universe: A Quantum Computer Scientist Takes on the Cosmos, Vintage, 2007.

Mandelbrot, B.B.
The Fractal Geometry of Nature, Macmillan, 1983.

Maor, E.
To Infinity and Beyond: A Cultural History of the Infinite, Princeton University Press, 1991.

McFadden, J. and Al-Khalili, J.
Life on the Edge, Broadway Books, 2016.

Nahin, P.J.
An Imaginary Tale: The Story of $\sqrt{-1}$, Princeton University Press, 1998.

Penrose, R.
The Emperor's New Mind, Vintage, 1989.

Pickover, C.
Computers, Pattern, Chaos, and Beauty: Graphics from an Unseen World, Dover Publ., 2001.

Pickover, C.
The Math Book – From Pythagoras to the 57th Dimension, 250 Milestones in the History of Mathematics, Sterling Publ., 2009.

Rees, M.
Just Six Numbers, Basic Books, 2001.

Reid, C.
Hilbert, Springer-Verlag, 1970.

Rovelli, C.
Reality is Not What it Seems: The Journey to Quantum Gravity, Riverhead Books, 2018.

Schiff, J.L.
Cellular Automata – A Discrete View of the World, John Wiley & Sons, 2008.

Schiff, J.L.
The Most Interesting Galaxies in the Universe, IOP Concise Physics, 2018.

Stanley, K.O.O. and Lehman, J.
Why Greatness Cannot be Planned: The Myth of the Objective, Springer, 2015.

Tegmark, M.
Our Mathematical Universe – My Quest for the Ultimate Nature of Reality, A. Knopf, 2014.

Wallace, D.F.
Everything and More – A Compact History of ∞, Weidenfeld & Nicolson, 2003.

Wolfram, S.
Cellular Automata and Complexity, Westview, 1994.

Wolfram, S.
A New Kind of Science, Wolfram Media, 2002.

Index

A

Aaronson, Scott, 210, 212
Abbott, Edwin A., 27
A Beautiful Mind, 6
Abel sum, 64
Adams, John Couch, 178
Adams, Walter S., 234
Ad infinitum, 44, 53, 54, 61, 62, 64, 74–75, 129, 152, 160, 165
Aleph naught $\left(\aleph_0\right)$, 46
Alexandria, Egypt, 174
Alexandrov, Pavel Sergeyevich, 236
Algebra, 9, 37, 54, 56, 89, 108, 109, 133, 231, 276, 280
Algorithm, 30, 63, 121, 132–135, 255
Al-Khwārizmī, Muhammad bin Mūsā, 133
Al-Khalili, J., 206
Alpha Centauri, 215
Alternating geometric series, 63, 64, 264
Alternating Riemann zeta function, 270
AND gate, *see* Logic gates
Anderson, Carl, 195, 196
Andromeda galaxy, 179, 181, 215
Angular momentum, 210, 243, 244
Anthropic Principle, 14
Anti-de Sitter model, 25–27, 223, 233
Antimatter, 197
Anti-particle, 195, 197, 198
Ants, 143–147, 149
Apparent brightness, 283
Area, 24, 25, 29–32, 35–37, 55, 176, 177, 192, 203, 220, 221, 280
Argand, Jean-Robert, 83

Argument (of a complex number), 83, 157, 213, 240, 254, 255, 261
Aristarchus of Samos, 175
Aristotle, 3, 4, 41
Armistice, 214, 215
Arrow paradox, 42
Ars Magna (by Cardano), 77
Associative (property), 81, 90, 91, 238–241
Asteroid, 107–109, 111–113, 279
Asteroid Belt, 107–109
Astronomy, 60, 121, 173, 189, 200, 235
Atomic clock, 190, 227, 228, 232
Attractor, 153–155, 161
Augenstein, Bruno, 51
Average, *see* Mean (average)
Axiom, 8, 18–20, 50, 51, 205, 245, 247, 248
Axiom of Choice, 50, 51
Axis of rotation, 88, 266

B

Babylonians, 2, 3, 29, 77
Bacteria, 122, 143, 147–149, 219
Baez, John, 88, 91
Banach-Tarski paradox, 51
Barber of Seville, 8
Barrie, James M., 9
Base 2, 140, 209
Base 60, 2
Bassler Bonnie, 148
Beauty (mathematical), 38, 61, 63, 160, 195, 236
Beckenstein, Jacob, 192
Becker, Adam, 204

© Springer Nature Switzerland AG 2020
J. L. Schiff, *The Mathematical Universe*, Springer Praxis Books,
https://doi.org/10.1007/978-3-030-50649-0

Beckmann, Petr, 29
Bees, 143, 147, 149, 150
Bernoulli, Jacob, 71
Berry, Michael, 97
Bessel, Wilhelm
    Problem, 60
Bhaskara, 77
Bierce, Ambrose, 3
Big Bang, 43, 58, 179, 187, 193, 197, 199, 250
Biology, 20, 39, 103, 121, 132, 144, 152, 153, 206
Bioluminescence, 148
Birkhoff, George David, 208
Bit, 117, 209
Black hole
    stellar mass, 189
    supermassive, 73, 111, 188–190
Blackjack, 103–105
Bloch, Felix
    sphere, 210, 211
Blueshift, 200
Bohr, Niels, 202, 204
Bolyai, János, 22
Bombelli, Rafael, 78, 79, 82
Boolean Algebra, 280
Boole, George, 3, 280
Born Rule, 202
Bosons, 74, 244
Bousso, Raphael, 28
Bradley, James, 169
Brahe, Tycho, 176
Brahmagupta, 77
Breeding, 134, 150
Bridges, Douglas, 6
Brooks, Robert, 159
Broom Bridge, 86, 87
Brough, Michael, 134
Brouwer, L.E.J.
    fixed-point theorem, 6
Brown, Robert, 103
Brownian Motion, 103
Buckingham, David, 129
Bulb (of Mandelbrot set), 161, 162
Butterfly effect, 153, 156
Butterworth, Brian, 9
Byte, 209

C
Calculus, 58, 59
Cantor, Georg
    diagonal method, 246, 260
Carbon-14 (C-14), 106
Cardano, Gerolomo, 77, 82
Cardinal, 46–48, 50–52, 248

Cardinal arithmetic, 47
Cardinality, 46–48, 50, 259, 261–262
Carroll, Sean, 203
Carter, Brandon, 14
Casimir Effect, 66
Cassini spacecraft, 226, 234
Catenary, 105
Cayley, Arthur
    numbers, 90
Cellular Automata (CA)
    principles, 122, 133, 137
Cepheids, 180, 184, 283–284
Ceres, 107
Cesàro sum, 64, 65, 265
Cesium (caesium), 227
Chaos, 98, 113–116, 119–167
Checkerboard, 122, 133, 134, 247
Chelyabinsk meteoroid, 111, 112, 279
Chemistry, 20, 32, 142, 144, 152, 210, 242
Choice function, 50, 51
Chou, James Chin-Wen, 227
Chromosomes, 133, 134
Chudnovsky algorithm, 30
Circle, 9, 18, 21–26, 28–32, 35–38, 56, 60, 83, 84,
        93, 99, 102, 135, 142, 157, 161, 169,
        176, 240–242, 244, 274
Circle group, 241, 242
Circumference of Earth, 174, 175
Clausius, Rudolf, 115
Cloning, 133
Closed (group), 5, 6, 35, 37, 99, 114, 116, 118,
        186, 191, 238–240
Closed Universe, 114
Coastline, 165, 166
Cohen, Paul, 50
Collisions, 15, 33–35, 107, 113, 195, 198, 199
Cologne Cathedral, 142, 143
Commutative (property), 81, 91, 240
Complex
    analysis, 11, 82
    number, 76, 80–89, 91, 93–94, 157, 160, 161,
        202, 206, 208, 210, 239, 240, 271
    plane, 81–84, 92, 94–96, 98, 206, 240, 270, 272
    variables, 76, 91, 92, 271
Complex adaptive system, 142
Composite number, 254, 255
Compound interest, 68, 71
Compte, Auguste, 178
Compton, Arthur, 201
Computer, 6, 30, 32, 34, 52, 56–58, 63, 67, 69, 71,
        88, 97, 101–103, 117, 118, 121,
        123–125, 127, 130, 133, 140, 153, 158,
        159, 166, 167, 189, 195, 206, 209–212,
        215, 228, 246–248, 282

Consciousness, 142, 144
Conservation laws, 243, 244
Continuous group, 241, 242
Continuum, 26, 42, 48, 222, 248, 250
    Hypothesis (CH), 50, 51, 246, 248
Convergent (series), 54, 60
Conway, John Horton, 121, 127, 129, 205
Copenhagen interpretation, 202, 204, 206
Copernicus, Nicolaus, 175, 176
Cosine (*cos*), 52, 63, 158
Cosine function, 52, 62
Cosmic ray, 106, 195, 198
Cosmological constant (Λ), 13, 223
Cosmology, 14, 28, 173, 187, 290
Coulomb's Law, 220
Countable, 46, 260
Cox, Brian, 99
Creator, 14, 121, 124, 128, 208
Critical
    line, 96–98, 272
    strip, 270
Critical density, 185–187, 249
Crystallography, 242
Cygnus X-1, 188, 189
Cylinder, 26, 27, 125, 187

D
Dali, Salvador, 152
Dark energy, 187, 223, 248–250
Dark matter, 187, 244
Darwin, Charles, 132–136
Davies, Paul, 14
De Broglie-Bohm Theory, 206
De Broglie, Louis, 202
Decay
    constant, 108, 276–278
    exponential, 105, 106, 108, 109, 276
Degree, 9, 15, 18, 21, 24, 28, 62, 116–118, 120,
    176
De Morgan, Augustus, 3
Density
    parameter, 13, 185, 186
De Pretto, Olinto, 217
Derbyshire, John, 97
De Sitter, Willem, 25, 223
Diffusion, 103, 122
Dinosaurs, 16, 113
Dirac, Paul Adrien Maurice (P.A.M.), 195, 196,
    250
Dirichlet Problem, 100, 135
Discontinuity, 42, 100
Discrete
    group, 241
    time, 43, 52, 121, 123, 124

Disk, 6, 15, 22–27, 160, 161, 189, 224, 257, 258
Dispersion, 73, 190
Divergent (series), 56, 59, 61, 65
DNA, 15, 119
Dot/Inner product, 209
Drone bee, 149, 150, 152
D-Wave, 212
Dynamical System, 98, 129, 141, 152–155, 157,
    158
Dyson, Sir Frank Watson, 224, 226

E
Earth, 14–16, 28, 78, 106–109, 111–114, 149,
    164, 168, 169, 173–176, 180, 182, 184,
    189, 191, 194, 195, 204, 215, 216,
    218–221, 226–229, 232–236, 245, 278
Eccentricity, 176
Economics, 6, 103, 113, 121, 126
Eddington, Sir Arthur, 114, 224, 225
Egyptians, 29
Einstein, Albert, 28, 41, 103, 168, 188, 201, 203,
    205, 213, 214, 217, 219, 223–226, 231,
    232, 243, 253
Electromagnetic
    forces, 91
    radiation, 169, 213, 220
    spectrum, 169, 170
    waves, 169, 171, 172, 213
Electron, 12, 188, 195–198, 200–204, 210, 244,
    248, 249
Element (set), 6–8, 13, 45–47, 49, 238–240
Electron volt, 200
Elementary particles, 91, 198, 206
Elements (Euclid's), 11, 17, 174
Eliot, T.S., 113, 114, 194
Ellipse, 176, 177
Empty set, 7
Empty space, 13, 169, 198, 248
Encryption, 242
Energy, 13, 33, 42, 75, 85, 97, 98, 111, 112, 115,
    116, 118, 152, 169, 171, 187, 195–201,
    205, 206, 217, 218, 220, 221, 223, 224,
    227, 234, 243, 244, 248–250, 279
Engineering, 32, 52
Entropy
    information, 116–119, 192
Epstein, R.L., 6
Equilibrium state, 135
Eratosthenes, 173–175
Erdös, Paul, 11, 18
Ernst Zermelo, 8, 50
Erzberger, Matthias, 215
Escape time, 158, 159
Escher, Maurits Cornelius (M.C.), 24, 25

E Silva, Tomás Oliveira, 247
Estimating the Maximum (EMX), 247, 248
Euclidean
    distance, 23, 258
    geometry, 18, 20
    plane, 18, 78, 86
    triangle, 20, 24, 25
Euclid of Alexandria
    Fifth Postulate, 18–23
Euler, Leonhard
    formula, 84, 85, 97, 273
    identity, 82–85
Event horizon, 189, 191, 192
Everett III, Hugh, 203
Expansion of the Universe, 13, 179, 180, 187, 223, 224, 248, 249, 289–290
Expansion of Universe, 249
Exponential
    decay, 105–106, 108, 109, 276
    e, 30, 105, 109
    function, 67, 69, 71–73, 84, 105, 158
    growth, 105, 109
Extinction, 16, 153

F
Fatou, Pierre
    set, 158
Ferguson, Thomas, 71
Fermions, 244
Feynman, Richard, 20, 38, 85, 120, 142, 199
Fibonacci
    retracements, 152
    sequence, 150, 152
Field equations (Einstein), 25, 185, 188, 249
Fine-tuning, 12–16
Fitness test, 133, 134, 247
Fixed-point theorem
    Brouwer, 6
    Kakutani, 6, 101
Flat Earth Society, 173
Flat Universe, 185
Flatland, 27
Foch, Marshall Ferdinand, 215
Foot-candle, 184
Fourier, Joseph
    analysis, 52
    series, 52
Fractal, 158, 160, 162–167
Fraenkel, Abraham, 8, 50
Fredkin, Edward, 121, 211
Free Will Theorem, 205
Frequency
    of occurrence, 72–74, 110, 112

shift, 226, 234
    of a wave, 111, 171, 200, 234
Friedmann, Alexander, 185, 223
Friedmann-Lemaître-Robertson-Walker (FLRW)
        model, 224
Fundamental frequency, 37
Fundamental theorem of arithmetic, 254
Furey, Cohl, 91

G
Galaxy, 14, 111, 114, 164, 176, 179–184, 187, 188, 190–193, 197, 215, 224, 234, 236, 249, 283–285, 289
Galilei, Galileo, 168
Galle, Johann Gottfried, 178
Galperin, Gregory, 32, 34, 35
Gambler, 104, 105
Gambling, 69, 105
Game of Life, 121, 123, 127–130, 138, 144, 154
Game Theory, 121
Gamma function, 271
Gauss, Carl Friedrich, 22
Gell-Mann, Murray, 38, 198
General relativity, *see* Relativity
Genes, 9, 134
Genetic algorithm, 132, 134
Geometric series, 54, 55, 63, 263, 264
Glider
    gun, 129, 130
Global positioning system (GPS), 11, 214, 218, 228, 232, 233
God, 2, 9, 11, 14, 41, 76, 85, 151, 194, 196, 205
Gödel, Kurt
    incompleteness theorem, 246
Goldbach, Christian
    conjecture, 246, 247
Golden ratio (mean), 151
Goldilocks, 14, 15, 186
Google, 211, 212
Gosper, Bill, 129
Göttingen University, 242
Grandi, Guido
    series, 64, 265
Graves, John T., 90
Gravitational lensing, 192, 193
Gravitational redshift, 234, 236
Gravity, 13, 14, 26–28, 107, 114, 178, 188–190, 192, 193, 214, 219–229, 233–235, 249, 250
Greene, Brian, 12, 14, 41
Gregory, James, 56
Grewal, Damanveer S., 15
Groups, 13, 46, 72, 202, 236–244, 247

H

Habitable zone, 15
Half-life, 106, 108, 198
Halting Problem, 246–247
Hamilton, William Rowan, 78–80, 86–88, 90
Hamming, Richard, 17, 76
Haraguchi, Akira, 29
Hardy, Godfrey Harold (G.H.), 11, 66, 97
Harmonic
    function, 100, 101
    series, 57–60, 93, 94
Hasenöhrl, Friedrich, 217
Hawking, Stephen, 192
Heisenberg, Werner, 202, 204, 248–250
Helfgott, Harald, 247
Herschel, William, 178
Hertz, Heinrich
    Hertz (Hz cycles per second), 171
Higgs particle, 74, 198, 199
Hilbert, David
    space, 202, 208–210, 212
H0liCOW, 187
Hooft, Gerard 't, 26, 192
Holographic Principle, 26, 27, 100, 192, 251
Hubble, Edwin
    constant, 180, 181, 185–187, 191
    parameter, 181, 185, 290
    Space Telescope, 187, 190, 283
Hydrogen atom (mass), 185
Hyperbolic
    distance, 24, 25
    Geometry, 22, 23, 25, 26, 28, 256–258
    paraboloid, 21
    triangle, 21, 24, 25
Hyperboloid, 22, 256–258

I

IBM
    Q Experience, 212
Ice, 115, 116, 197
Identity (of a group), 239
Imaginary
    number (i), 11, 76, 80–82, 85–87, 90, 195
    part, 81, 96
Inelastic scattering, 198
Infectious disease model (SIR), 131–132
Inference rule, 4
Infinite
    product, 74, 75, 273
    series, 30, 44, 52–57, 62–65, 67, 69, 93, 263, 265, 271
    set, 5, 7, 26, 44–47, 51, 74
    space, 24, 187, 206–209, 266
    sum, 52, 53, 59, 64, 93, 94, 208, 273

Infinity, 5, 22, 24–26, 39–75, 93, 94, 130, 157–159, 162, 233, 258, 271
Information entrophy, *see* Entropy
Infrared, 169
Ingenhousz, Jan, 103
Integers
    even ($\mathcal{E}$), 44–47, 271
    negative, 47, 95
    odd ($\mathcal{O}$), 45, 46
    positive ($\mathcal{P}$), 4, 44, 47, 60, 254, 259, 260, 271
Intensity (energy), 220
Intrinsic brightness, 184, 283
Inverse (of a group), 239
Iodine-131, 105, 106
IOK-1 galaxy, 179
Irrational number, 48
Isoperimetric Inequality, 35–38
Iterates, 153, 157, 158, 161, 162
Iterations, 132, 141, 153, 154, 156–161
Iwao, Emma Haruka, 30

J

Jordan, Pascual, 204
Joules, 115, 116, 279
Julia, Gaston
    set (filled-in), 158, 159
Jupiter, 107, 109, 168, 176, 219

K

Kakutani, Shizuo
    fixed-point theorem, 6, 101
Kauffman, Stuart (S.A.), 146
Keating, Jonathan, 97
Kelvin, 114, 116
Kepler, Johannes
    Laws, 176, 177
Kerr, Roy
    black hole, 188, 189, 191
Klein, Felix, 242, 243
Kochen, Simon, 205
Kovalevskaya, Sofia, 1
Kronecker, Leopold, 9, 47, 166
K/T (boundary), 113

L

L'Algebra (by Rafael Bombelli), 78
Langton, Christopher, 144
    Ant, 144, 146
    Highway, 146
Laplace, Pierre-Simon, 188, 190
Large Hadron Collider, 199

Large Magellanic Cloud (LMC), 283–285
Lathe, Richard, 15
Law of excluded middle, 4, 5, 253, 260
Law of Universal Gravitation, 164, 176, 220
Laws of Motion, 176, 236
Lead (Pb-206), 106
Le Corbusier (Charles-Édouard Jeanneret), 152
Le Verrier, Urbain, 178
Leavitt, Henrietta Swan, 180, 184, 283
Lebowitz, Fran, 9
Leibniz, Gottfried Wilhelm
    series, 56, 57
Lemaître, Georges, 223, 224
Lennox, John C., 14
Lewin, Roger, 142
Light
    clock, 215, 227, 229–231, 286, 287
    speed (c), 85, 169, 171–173, 189, 213–215,
        228, 231, 233, 234 (see Speed
        of light (c))
Light-year, 39, 178–180, 182, 189–191, 215, 234,
        235, 283, 284
Limit, 25, 37, 42, 50, 54, 55, 64, 68–70, 74, 164,
        165, 209, 213, 245, 264
Littlewood, John Edensor (J.E.), 66
Livio, Mario, 224
Lloyd, Seth, 210
Lobachevsky, Nikolai, 22, 28
Locality (Principle), 124, 130, 137
Logarithm, 24, 33, 58, 71, 85, 110, 117,
        164, 278
Logic gates
    AND gate, 281
    NOT gate, 281
    OR gate, 281
Logistic equation, 153, 160, 161
Log-log plot, 112
Lookback time, 191
Loop Quantum Gravity (LQG), 124, 250
Lord Rayleigh (John William Strutt), 37
Los Alamos, 32, 121
ℓ^2 space, 202, 208–210, 212
Lynds, Peter, 42

M
Mādhava-Leibniz series, 56
Magnetic moment, 204
Mahler's Fifth Symphony, 1
Main cardioid, 160, 161
Maldacena, Juan, 26–28
Mandelbrot, Benoit
    set, 154, 159–162, 165, 224

Manhattan Project, 121
Many-Worlds (interpretation), 203
Maor, Eli, 44
Mass, 15, 32–34, 73, 85, 105, 107, 114, 164,
        185–190, 195, 198–202, 217, 218, 220,
        224, 233–236, 243, 244, 248, 249, 277,
        279
Mass-energy density, 185
Mass spectrometer, 277
Matelski, J. Peter, 159
Mathematical Universe Hypothesis (MUH), 13
Matlab, 57, 63
May, Robert M., 153
McFadden, J., 206
Mean (average), 44, 72, 73, 100, 135, 190
Mean value property, 100, 135
Meena, Rajveer, 29
Megaparsec, 180
Meteorites, 107–109, 111, 113, 276–279
Meteoroids, 111, 112, 279
M87 galaxy, 190
Michell, John, 188
Michelson-Morley experiment, 214
Microwave, 169, 171–173, 186, 187, 213
Milky Way galaxy, 181, 183, 190
Minkowski, Hermann
    Spacetime, 25, 222, 233, 256–258
Mirror reflection, 237
Misiurewicz points, 162
Möbius strip, 187
Modulus, 82, 83, 92, 157, 159, 240
Modus ponens, 4
Modus tollens, 4
Momentum
    linear, 243, 244
Monte Carlo Method, 32
Moon, 15, 107, 168, 176, 204, 224
Moore, Edward F.
    neighborhood, 124, 131
Morgenstern, Oskar, 121
Morrison, Grant, 39
Mozart, 152
M-theory, 233
Multiverse, 13, 14, 203
Mutation, 134

N
Nagasaki, 218
Nahin, Paul, 82
Nanometer, 66, 169, 200
Nash equilibrium, 6
Nash, John, 6, 101

Nature, 12, 41, 74, 100, 104, 111, 120, 121, 142, 144, 153, 173, 189, 195, 201, 211, 212, 218, 224, 226, 236, 237, 244, 246, 284
Negative curvature, 21
Negative integers, *see* Integers
Neighborhood-states, 133
Neptune, 178, 224
Neutrino, 197
Neutron, 12, 197, 198, 206
Newton, Isaac, 176, 185, 190, 219, 220, 222, 224, 236
Nitrogen-14, 106
Noether, Emmy, 242–244
Noether's theorem, 243, 244
Nonassociative, 91
Norm (of a vector), 207, 209
Normal distribution, 73, 74
Normal families, 1
NOT gate, *see* Logic gates
Number module, 9

O

Octonians, 89–91
Odlyzko, Andrew, 97
One-dimensional, 122, 137–142, 163, 165, 189, 206, 222, 233
One dimensional automata, 137
One-to-one correspondence, 45, 48–50, 259, 260, 262
Open Universe, 186
Optimization, 149
Orbit, 15, 60, 98, 153, 156, 161, 162, 164, 168, 175–178, 182, 190, 215, 221, 222, 228, 229
Ordered pair, 78–82, 86
Ordinary matter, 187, 197, 223
OR gate, *see* Logic gates

P

Parallax, 182–184, 283
Parallel Postulate, 18, 21, 22
Pareto Principle (80/20 Rule), 113
Parsec, 180, 182, 191, 283
Parthenogenesis, 143
Partial sum, 53–56, 59, 60, 64, 65, 69, 74, 263–265
Pascal's triangle, 138
Peirce, Charles Sanders, 3, 47
Penrose, Roger, 11
Perimeter, 35–37
Period, 15, 27, 43, 58, 71, 129, 131, 144–147, 154, 161, 162, 176, 184, 208, 211, 231, 243, 247, 283, 284, 286

Period four
    4-cycle, 154, 155, 161
Period three
    3-cycle, 154
Period two
    2-cycle, 146, 154, 161
Periodicity, 161, 162
Perlmutter, Saul, 249
Petermann, André, 198
Photoelectric effect, 200, 201
Photon, 43, 200, 201, 244, 249
Phyllotaxis, 152
Physics, 12–14, 20, 26–28, 32, 33, 37–40, 42, 43, 51, 56, 60, 74, 85, 91, 98, 103, 120, 121, 142, 144, 152, 169, 171, 189, 192, 196–199, 204, 205, 210, 211, 214, 217, 236, 237, 242–244, 248, 249
Pi/π, 21, 24, 25, 28–32, 34, 36, 37, 48, 56, 57, 60, 63, 67, 69, 75, 84, 85, 93, 94, 99, 105, 169, 189, 223, 241, 248, 271–273
Pickover, Clifford, 85
Planck, Max
    length, 42, 250
    Space Telescope, 186, 249
    time, 43
Planetesimals, 15, 107
Plato
    Platonism, 10
    Republic, 251
Playfair, John
    axiom, 19, 20
Pletser, Vladimir, 152
Poincaré, Henri
    disk model, 22, 23
Polchinski, Joseph, 66
Pollock, Jackson, 118, 119
Pólya, George, 98
Polynomial, 61, 62, 80, 89, 90, 95
Popper, Daniel, 234, 235
Positive integers, *see* Integers
Positron, 195–199
Positron Emission Tomography (PET), 197
Postulates, 16–23, 203, 213, 214
Power laws 109–111, 113, 164
Power series, 61–63, 67, 85
Precession, 221, 222
Preskill, John, 211
Prime (numbers), 4, 5, 59, 97, 246, 247, 253–255, 273
Principal value, 85
Principia Mathematica (Newton's), 176, 220
Principle of complementarity, 201
Principle of least action, 187, 244
Probability, 70, 101, 102, 117, 118, 131, 202–206, 210, 233, 248

Proof by contradiction, 5, 6, 255
Proton, 12, 197–199, 206
P-series, 93, 94, 273
Pythagoras
    theorem, 10, 231
Python, 274

Q

Quadratic formula, 77, 95, 151
Quantum
    biology, 206
    computer, 209–212
    Mechanics, 10, 11, 26, 42, 52, 75, 82, 88, 97,
        119, 121, 195, 196, 199, 203–206, 208,
        209, 233, 242, 248, 250
    supremacy, 211, 212
    tunneling, 206
Quark, 51, 198, 199
Quaternions, 86–91, 266–269
Qubit, 117, 209–212
Queen bee, 143, 149
Quorum sensing, 147–149

R

Radial velocity, 180–182, 289
Radian, 21, 24, 62, 63, 83–85, 240, 241
Radiation, 114, 169, 187, 213, 220, 234–235, 249
Radioactive
    decay, 105, 106, 108, 197, 198, 277
Ramanujan, Srinivasa
    summation, 65, 66
Random
    walk, 101–105, 135, 274–275
Rational (number), 12, 48
Real number, 6, 40, 42, 48–52, 61, 77–82, 85, 86,
        90–94, 97, 98, 157, 202, 206, 239, 240,
        259–260, 269, 271–273
Real part, 81, 94–96, 273
Recurrence relation, 152
Redshift, 180–182, 190, 191, 193, 200, 224,
        234–236
Reductio ad absurdum, 5, 246
Rees, Martin, Baron Rees of Ludlow, 13, 14
Relational reality, 13
Relativity
    General, 218, 222, 223, 225, 226, 250
    Special, 88, 195, 217, 250, 258
Repeller, 153
Riemann, Bernhard
    Hypothesis, 91–98, 208, 245
    Zeta function, 61, 66, 95–98, 270–273
Riemannian Geometry, 21

Riess, Adam G, 183, 249
Roberts, Cokie, 9
Rogers, Will, 9
Rømer, Ole, 168, 169
Roots, 2, 11, 78, 95
Rota, Gian-Carlo, xv
Roulette, 69, 70
Rovelli, Carlo, 42, 250
RSA encryption, 242
Rubidium-87 (Rb-87), 108, 276, 278
Rule 1 (CA), 137–139
Rule 18 (CA), 139–141, 156, 162, 164
Rule 90, 139
Russell, Bertrand
    Paradox, 7, 8, 20

S

Saddle, 20–22, 24
Sagittarius A*, 222
Scale-invariance, 164
Scaling factor, 164, 289
Schelling, Thomas, 126
Schmidt, Brian P., 249
Schrödinger, Erwin
    equation, 195
Schwarzchild, Karl
    radius, 188, 189
Second Law of Thermodynamics, 114, 116, 120,
        191
Secretary/Marriage problem, 70, 71
Seed value, 157, 159, 162
Self-organizing system, 143
Self-similarity, 140, 158, 162, 165–167
Semi-major axis, 176
Semi-minor axis, 176
Sensitivity to initial conditions, 141, 156
Set, 5–9, 11–13, 26, 44–48, 50, 51, 74, 80, 81, 83,
        84, 130, 138, 154, 158–162, 165, 206,
        223, 224, 228, 239–241, 246, 262
Set Theory, 8, 50, 51, 246, 248
Shakespeare, 245
Shannon, Claude, 116–118
SH0ES, 187
Sierpinski triangle, 140, 141, 164
Signal processing, 52, 82, 123
Sikhote-Alin, 113
Similarity/Fractal dimension, 164–166
Simultaneity, 214, 216
Sine, 63, 158, 271
Sine function, 52, 62
Singularity, 40, 91, 92, 94, 95, 189, 250
SIR model, *see* Infectious disease model (SIR)
Slipher, Vesto, 180, 224

Smolin, Lee, 66
Snowflake, 237, 239, 240
Socrates, 8, 251
Solar Nebula, 107, 278
Solar System, 15, 107–109, 114, 176, 178,
        276–278
Spacetime, *see* Minkowski, Hermann, Spacetime
Spatial translation, 236, 242, 244
Special Relativity, *see* Relativity
Speed of light (c), 85, 169, 171–173, 189,
        213–215, 228, 231, 233, 234
Sphere, 20, 21, 26, 37–39, 89, 100, 176, 186, 188,
        210, 211, 220, 221, 228, 250, 258, 269,
        279
Spin, 70, 189, 204, 210, 244
SPT0615-JD galaxy, 193
Square summable, 208
Stadia, 174, 175
Standard deviation
    5-sigma event, 74
Standard Model of Physics, 91
Standing wave, 172, 173
Star, 14, 15, 28, 114, 175, 178, 180, 182–184, 187,
        189, 190, 206, 215, 220, 222, 224,
        233–236, 249, 283, 284
Starobinsky, Alexei, 187
State function, 202
Statistics, 32, 72, 73
Steady-state, 100, 135
Stellar mass black hole, *see* Black hole
Straight line, 9, 18–23, 48, 82, 109–111, 140, 141,
        164, 165, 175, 181, 190, 214, 243, 244,
        257, 277, 278, 283
String Theory, 36, 66, 98, 233, 244, 250
Strontium-87 (Sr-87), 108, 109, 276, 277
Styer, Daniel, 204
Sun, 14, 109, 114, 118, 169, 174–178, 182, 184,
        188–190, 192, 197, 215, 218, 221, 222,
        224–227, 234, 235
Supermassive black hole, *see* Black hole
Supernova, 184, 249
Superposition (of states), 210, 212
Supersymmetry (SUSY), 244
Susskind, Leonard, 26, 192
Sycamore computer chip, 212
Syene, Egypt, 174
Symmetry, 96, 147, 152, 236–244
Synchronicity (Principle), 124, 130, 137

T
Tegmark, Max, 12, 13, 203
Teller, Edward, 121
Temperature, 99–103, 114–116, 125, 135, 136,
        143, 178, 187, 274

Termites, 143–144, 147
Tessellation, 24, 25
Theia, 15
Theory of Evolution, 121
Theory of Relativity, *see* Relativity
Thermal radiation, 114
Three-dimensional, 18, 22, 25, 26, 37, 38, 42, 51,
        86, 88, 89, 92, 163, 187, 206–209, 220,
        222, 262, 266, 269
Time dilation, 219, 226–228, 231–233, 286
Time translation, 236, 244
Tip of the Red Giant Branch (TRGB), 184, 187
Topology, 187
Torus, 125, 187, 240
Transfinite
    arithmetic, 47, 48, 50
    cardinal, 50, 51
Transistor, 117, 123, 206, 209
Triangle Postulate, 21
Trigonometry, 62, 82, 158, 182, 183, 268
Trilateration, 228
Trivial zeros, 95, 271
Truth tables, 280
Tukey, John, 117
Tunguska event, 111
Turk, Greg, 144
Twin paradox, 231
Two-dimensional
    cellular automata, 122, 137
    plane, 78, 80, 82, 84, 86, 206
Type Ia supernovae, 184

U
Ulam, Stanisław, 32, 121
Ultraviolet, 169
Uncertainty Principle, 248
Uncountable, 259–260
Uniformity Principle, 124
Universe, 2, 7, 10–15, 18, 22, 24, 26–29, 35–37,
        39, 41, 51, 58–66, 76, 82, 98–119,
        121–124, 130, 133, 137, 142, 148, 160,
        162–169, 173, 175–182, 184–187,
        192–199, 202–205, 209–211, 213, 214,
        217, 220, 223, 224, 227, 231, 236, 237,
        242–252, 254, 289–290
Unreasonable effectiveness of mathematics, 12, 76
Uranium (U-238), 106, 195, 198
Uranus, 60, 178

V
Vacuum energy, 187, 248
Variable stars, 180, 184, 283
Veltman, Martinus, 192

Verhulst, Pierre, 153
Vesta, 107, 111
Viberio harveyi, 148
Voltaire (François-Marie Arouet), 41
von Hardenberg, Georg Friedrich Philipp, 1, 85
von Neumann, John
 neighborhood, 124
von Neumann neighborhood, 124
Vote/Majority (Rule), 125

W
Wallis, John, 75
Walsh, Malcolm
 highway, 147
Watkins, Ron, 69
Wave
 function, 202, 203, 205, 206, 210
 length, 169, 171–173, 182, 184, 200, 202, 234,
  236, 249
 Mechanics, 202
 radio, 169–171, 213, 226, 234
 standing, 172, 173
Wave-particle duality, 202
Wessel, Casper, 83
Weyl, Hermann, 5, 38
Wheeler, John, 188, 224
White dwarf
 40 Eridani B, 234–236
Whittaker-Shannon interpolation formula, 52

Whole number, 7, 44, 46, 48, 51, 54, 58, 153
Wiener, Norbert, 116
Wigner, Eugene, 12, 20, 242
Wilkenson Microwave Anisotropy Probe
 (WMAP), 186, 187
Wolfram, Stephen, 121
Work, 115
Worker bees, 143, 149

X
X-ray, 169, 171, 189

Y
Y-cruncher, 30
Yee, Alexander, 30
Ytterbium, 227

Z
Zel'dovich, Yakov, 187
Zeno of Elea
 Paradox, 41–52, 55
Zermelo-Fraenkel set theory, 8, 50
Zero (of a function), 95, 97, 98, 133, 271
Zero curvature, 21
Zero-point energy, *see* Vacuum energy
Zweig, George, 198